D1429296

Modeling and Simulation in Science, Engineering and Technology

Series Editor
Nicola Bellomo
Politecnico di Torino
Italy

Advisory Editorial Board

Nicola Bellomo

Modeling Complex Living Systems

A Kinetic Theory and Stochastic Game Approach

Birkhäuser
Boston • Basel • Berlin

Nicola Bellomo
Dipartimento di Matematica
Politecnico di Torino
Corso Duca degli Abruzzi 24
10129 Torino
Italy

Mathematics Subject Classification: 00A71, 47J35, 74A25, 93A30

Library of Congress Control Number: 2007934430

ISBN-13: 978-0-8176-4510-6   e-ISBN-13: 978-0-8176-4600-4

Printed on acid-free paper.

9 8 7 6 5 4 3 2 1

www.birkhauser.com   (Int/SB)

To Fiorella and Piero

Contents

Preface

The subject of this book is the modeling of complex systems in the life sciences constituted by a large number of interacting entities called *active particles*. Their physical state includes, in addition to geometrical and mechanical variables, a variable called the *activity*, which characterizes the specific living system to be modeled. Interactions among particles not only modify the microscopic state, but may generate proliferative and/or destructive phenomena. The *aim* of the book is to develop mathematical methods and tools, even a new mathematics, for the modeling of living systems. The *background idea* is that the modeling of living systems requires technically complex mathematical methods, which may be substantially different from those used to deal with inert matter.

The first part of the book discusses methodological issues, namely the derivation of various general mathematical frameworks suitable to model particular systems of interest in the applied sciences. The second part presents the various models and applications.

The mathematical approach used in the book is based on mathematical kinetic theory for active particles, which leads to the derivation of evolution equations for a one-particle distribution function over the microscopic state. Two types of equations, to be regarded as a general mathematical framework for deriving the models, are derived corresponding to short and long range interactions.

This new kinetic theory, which can be applied to derive various models of practical interest in the life sciences, includes, as special cases, the classical models of kinetic theory, namely the Boltzmann and Vlasov equations. The difference, with respect to the classical theory, is that interactions follow stochastic rules technically related to the strategy developed by individuals belonging to living systems.

Various models and applications, derived within the mathematical framework, are presented in the second part of the book following a common style for all chapters: phenomenological interpretation of the physical system in view of the mathematical modeling; derivation of the mathematical model using methods of mathematical kinetic theory for active particles; simulations, parameter sensitivity analysis, and critical analysis of the model; and ideas to improve the existing models with special attention to applications.

Specifically, the following classes of models are studied: social competition related to individuals characterized by a microscopic state referred to their social collocation, modeling of vehicular traffic flow, immune competition for multicellular systems, and modeling of crowds and swarms with special attention to the analysis of the transition from normal to panic situations.

The complexity of the mathematical structures needed to deal with these systems increases from the first to the last class of models. For instance, models of social dynamics refer to spatially homogeneous systems, while traffic flow models need a space structure; immune competition models refer to a variable number of active particles due to the presence of proliferative and/or destructive events. Finally, models of crowds and swarms take into account a complex interaction space dynamics with rules which may consistently be modified by changes in the environment. The common characteristic for all these models is the attempt to describe collective behavior starting from microscopic dynamics.

The model selection has been influenced by the personal experience of the author. It is hoped that the reader interested in modeling living systems can generalize the approach to different fields of applications which may even produce further technical developments of the mathematical structures proposed in this book.

The chapters do not simply review the existing literature, but also propose new ideas and perspectives. The presentation mainly refers to modeling and simulations, and analytic aspects, for example, good position of mathematical problems and qualitative analysis of their solutions, are critically analyzed and brought to the attention of applied mathematicians as interesting research perspectives. The analysis of the models generates a variety of analytic and computational problems which are sufficiently complex to attract the intellectual energies of applied mathematician interested in the challenging perspective of modeling living systems.

Nicola Bellomo

1

From Scaling and Determinism to Kinetic Theory Representation

1.1 Scaling and Determinism

Systems of the real world are generally composed of several interacting elements. This implies that mathematical models can be designed at various observation and representation scales.

The *microscopic scale* corresponds to modeling, by mathematical equations, the evolution of a variable suitable to describe the physical state of each single object. An alternative to this approach can be developed if the system is constituted by a large number of elements and it is possible to obtain suitable local in space averages of their state in an elementary space volume ideally tending to zero. In this case, the modeling can be developed at a *macroscopic scale*, which refers to the evolution of locally averaged quantities, called *macroscopic variables*.

Different classes of equations correspond to these scalings. Generally, models designed at the microscopic scale are stated in terms of ordinary differential equations, while models at the macroscopic scale are generally stated in terms of partial differential equations. The modeling is developed within the framework of deterministic causality principles unless some external noise is added. This means that once a cause is given, the effect is deterministically identified.

Motivations to use the macroscopic scale instead of the microscopic one are also related to the practical objective of reducing complexity. For instance, when systems involve a large number of interacting elements, the number of equations of the model may be too large to be computationally tractable. Moreover, often only macroscopic quantities are of practical

1

interest so that it is convenient to deal with equations directly involving these variables.

A typical example is the flow of a fluid of several interacting particles. The microscopic scale corresponds to the motion of each particle using the framework of Newtonian mechanics. Particles move under mutual actions and under the influence of the external field. However, when the number of particles is sufficiently large and a sufficiently small volume still contains enough particles to allow the above-mentioned averaging process, macroscopic models can replace the microscopic description. Thus, the evolution in time and space of density, momentum, and energy, which are the macroscopic quantities that define the physical state of the fluid, are described by mathematical equations to model the system under consideration.

An alternative approach is the ***statistical (kinetic) representation***, where the state of the whole system is described by a suitable probability distribution over the microscopic state of the interacting elements, which is the dependent variable in kinetic models. Macroscopic observable quantities are computed by moments weighted by the distribution function.

This book deals with the modeling of large complex systems in applied sciences, with special attention to large systems of individuals whose dynamics follow rules determined by some organized, or even intelligent, ability. It is a rather different approach with respect to the traditional methods of mathematical kinetic theory, which refers to elements of inert matter, where interactions follow rules of classical or quantum mechanics.

In principle, kinetic theory can be applied to model systems belonging to living matter. In this case, one focuses on systems characterized by a high inner complexity: laws of classical or quantum mechanics cannot be straightforwardly applied, and the ability of interacting elements to organize their dynamics needs to be properly taken into account. In particular, interactions at the microscopic level are modeled according to stochastic rules, while interactions of classical particles follow deterministic rules.

Some books, among others those edited by Bellomo and Pulvirenti (2000), Degond, Pareschi, and Russo (2003), and Cercignani and Gabetta (2007), give an account of the surveys related to the application of the methods of kinetic theory to model complex systems. Examples and applications are reported in the book by Bellomo and Lo Schiavo (2000). More general aspects concerning the science of modeling are offered in e.g., the books by Lin and Segel (1988), Bellomo and Preziosi (1995), and Fowler (1997).

This book proposes a systematic approach to modeling large complex systems belonging to living matter by suitable developments of mathematical kinetic theory. This chapter provides a brief introduction to scaling, kinetic theory methods, and definitions which will be systematically used in other chapters. Moreover, a brief description of the aims and organization of the book is proposed. The contents of this chapter are as follows.

– Section 1.2 provides some historical and bibliographical indications on some classical models of mathematical kinetic theory to offer the reader the background necessary to understand the following chapters. This section briefly reports those models which refer to large systems of classical particles, as they are a fundamental reference for the models proposed in this book.

– Section 1.3 provides a brief introduction to discrete velocity models of classical kinetic theory, in view of some applications, such as vehicular traffic flow modeling, where discretization of the velocity space is useful to develop the specific modeling approach.

– Section 1.4 gives some preliminary definitions concerning the description of the microscopic state of the elements constituting a whole system, and its statistical description by suitable probability distributions. This section also introduces the concept of microscopic interactions and some preliminary ideas on modeling in applied sciences using kinetic theory methods. Moreover, some guidelines on the derivation of evolution equations are anticipated (this topic is exhaustively dealt with in Chapters 2–4).

– Section 1.5 describes the contents developed in the remaining chapters and defines the aims and organization of the book.

This book does not extensively report on the classical models and methods of mathematical kinetic theory. The interested reader is addressed to the pertinent literature for a deeper insight into methods of mathematical kinetic theory. Specifically, we refer to the book by Cercignani, Illner, and Pulvirenti (1994) for the derivation and application of the Boltzmann equation, to the books by Glassey (1995) and by Bellomo, Ed. (1995), as well as to the review by Villani (2004), for the qualitative analysis of mathematical problems. Moreover, the survey by Perthame (2004) provides a valuable presentation of several tools useful for dealing with kinetic equations. However, although models of mathematical kinetic theory are not described in detail, the Boltzmann and Vlasov equations are technically described and viewed as a particular case of a more general class of equations proposed in this book.

On the other hand, it is necessary to develop new mathematical methods and tools to model living systems, considering that these systems require a mathematical approach substantially different from those used to model inert matter. Not only may these tools be technically far more complex; a new mathematics may possibly need to be developed. This book proposes various ideas on this challenging topic.

The mathematical approach is developed in Chapters 2–4. The substantial difference with respect to classical kinetic theory is that interactions can be regarded as stochastic games, where the strategies developed by individuals plays an essential role in modeling the output of the interactions, once the input is properly assessed. Further, the reader does not need to possess special skills in methods of mathematical kinetic theory, because Chapters 2–4 offer a self-contained presentation of the mathematical methods used in this book.

1.2 Classical Models of Kinetic Theory

Mathematical models of kinetic theory have been developed to describe the dynamics of fluids from the viewpoint of statistical mechanics rather than that of the traditional approach of continuum mechanics.

A physical fluid is an assembly of disordered interacting particles free to move in all directions inside a space domain $\Omega \subset \mathbb{R}^3$ possibly equal to the whole space $\mathbb{R}^3$. Assuming that the position of each particle is correctly identified by the coordinates of its center of mass

$$\mathbf{x}_k, \qquad k = 1, \ldots, N, \tag{1.2.1}$$

the system can be reduced to a set of mass points in a fixed reference frame. For instance, this is the case when the shapes of the particles are spherically symmetric, and hence rotational degrees of freedom can be ignored.

When the domain Ω is bounded, the particles interact with its walls Ω_w. If Ω contains obstacles, that is subdomains $\Omega^* \subset \Omega$ which restrict the free motion, then the particles also interact with the walls of Ω^*. In some instances, namely for flows around space ships, the obstacles are objects scattered throughout the surrounding space.

It is generally believed in physics, and in statistical mechanics in particular, that understanding the properties of a fluid follows from a detailed knowledge of the physical state of each of its atoms or molecules. In most fluids of practical interest, such states evolve according to the laws of classical mechanics which, for a system of N particles, give the following set of ordinary differential equations:

$$\begin{cases} \dfrac{d\mathbf{x}_k}{dt} = \mathbf{v}_k, \\[2mm] \dfrac{d\mathbf{v}_k}{dt} = \dfrac{\mathbf{F}_k}{m_k} = \dfrac{\mathbf{f}_k}{m_k} + \dfrac{1}{m_k} \displaystyle\sum_{k'=1}^{N} \mathbf{f}_{k'k}, \end{cases} \tag{1.2.2}$$

with initial conditions

$$\mathbf{x}_k(0) = \mathbf{x}_{k0}, \qquad \mathbf{v}_k(0) = \mathbf{v}_{k0}, \qquad k = 1, \ldots, N, \tag{1.2.3}$$

where $\mathbf{F}_k$ is the force acting on the particle with mass m_k. $\mathbf{F}_k$ may be expressed as the superposition of an external field $\mathbf{f}_k$ and of the mutual force $\mathbf{f}_{k'k}$ acting on the k-particle due to the action of all other particles. In general, these forces are regular functions on the phase space, and $\mathbf{f}_{k'k}$ may be allowed to exhibit pole-discontinuities when the distance between particles is zero.

This approach prescribes that the mathematical problems stated by Eqs. (1.2.2)–(1.2.3) can be solved, and that the macroscopic properties of the fluid are obtained as averages involving the microscopic information contained in such a solution. However, it is very hard, and maybe not even possible, to implement such a program, unless suitable simplifications are introduced. Indeed, unavoidable inaccuracies in the knowledge of the initial conditions, the large value of N, and the mathematical complexity result in the impossibility of retrieving and manipulating all the microscopic information which can be supplied by (1.2.2)–(1.2.3) and contained in $\{\mathbf{x}_k, \mathbf{v}_k\}$ for $k = 1, \ldots, N$.

The practical interest consists in extracting from (1.2.2)–(1.2.3) information sufficient to compute the time and space evolution of a restricted number of macroscopic observable quantities such as number (or mass) density, mass velocity, temperature, and stress tensor. However, recovering the macroscopic observable quantities using the solution of Eq. (1.2.2) is an almost impossible task. This is due not only to the initial difficulty in dealing with a large system of ordinary differential equations, but also to the task of afterwards computing the averages which correctly define the macroscopic quantities.

For instance, the mean mass density $E(\rho)$ should be obtained, for a system of identical particles, by examining the ratio

$$E(\rho) = m \, \frac{\Delta n}{\Delta \mathbf{x}}, \tag{1.2.4}$$

when the volume $\Delta \mathbf{x}$ tends to zero and the number of particles Δn remains sufficiently large. Fluctuations cannot be avoided. Additional difficulties are related to the computation of the other macroscopic variables.

A different possible approach is the one of fluid dynamics; see, e.g., Truesdell and Rajagopal (2002) and Romano, Marasco, and Lancellotta (2005). It consists in deriving the evolution equations, related to the above macroscopic observable quantities, under several strong assumptions, including the hypothesis of continuity of matter, which requires that the ratio between the mass contained in a small volume and the volume itself is finite when the volume tends to zero. This is a good approximation of reality only if the mean distance between pairs of particles is small with respect to the characteristic lengths of the system, for instance the typical length of Ω or of Ω^*. Conversely, if the intermolecular distances are of the same order of such lengths, the continuum assumption is no longer valid, and a discrepancy is expected between the description of continuum fluid dynamics and that furnished by microscopic models, i.e., by Eqs. (1.2.2). Therefore, considering the inconsistency of using the equations of continuum fluid dynamics, and the complexity of dealing with particle dynamics, a different model is useful.

The idea (or paradigm) by Ludwig Boltzmann was to introduce the one-particle distribution function

$$f = f(t, \mathbf{x}, \mathbf{v}) : \quad \mathbb{R}_+ \times \Omega \times \mathbb{R}^3 \to \mathbb{R}_+, \quad (1.2.5)$$

and to derive an evolution equation for the distribution. ($\mathbb{R}_+$ means, here and afterwards, the set $\{\xi \in \mathbb{R}, \xi \geq 0\}$.)

If f is integrable at least locally, $f(t, \mathbf{x}, \mathbf{v}) \, d\mathbf{v} \, d\mathbf{x}$ defines the number of particles which at the time t are in the phase space volume $[\mathbf{x}, \mathbf{x} + d\mathbf{x}] \times [\mathbf{v}, \mathbf{v} + d\mathbf{v}]$. Moreover, if f is known such that $\mathbf{v}f$ as well as $v^2 f$ are integrable, the macroscopic observable quantities can be computed as expectation values of the corresponding microscopic functions. In particular,

$$\rho(t, \mathbf{x}) = m \, n(t, \mathbf{x}) = m \int_{\mathbb{R}^3} f(t, \mathbf{x}, \mathbf{v}) \, d\mathbf{v}, \quad (1.2.6)$$

where m is the mass of the particle (for a system of equal particles), and

$$\mathbf{U}(t, \mathbf{x}) = \frac{1}{n(t, \mathbf{x})} \int_{\mathbb{R}^3} \mathbf{v} f(t, \mathbf{x}, \mathbf{v}) \, d\mathbf{v}, \quad (1.2.7)$$

are, respectively, the mass density and the mass velocity, while the mean translational energy is given by

$$\mathcal{E}(t, \mathbf{x}) = \frac{1}{3(k/m)n(t, \mathbf{x})} \int_{\mathbb{R}^3} [\mathbf{v} - \mathbf{U}]^2 f(t, \mathbf{x}, \mathbf{v}) \, d\mathbf{v}, \quad (1.2.8)$$

where k is the Boltzmann constant. The mean energy can be related to the temperature by suitable assumptions on the thermodynamics of the system.

Global quantities are obtained by integration over the space variable, for instance the total number of particles is

$$N = \int_{\mathbb{R}^3 \times \mathbb{R}^3} f(t, \mathbf{x}, \mathbf{v}) \, d\mathbf{x} \, d\mathbf{v}. \quad (1.2.9)$$

Actually, one has to accept that the above kinetic type of modeling provides only an approximation of physical reality. For instance, the state of an N-particle gas is statistically described by the N-particle distribution function

$$f_N = f_N(t, \mathbf{x}_1, \mathbf{v}_1, \ldots, \mathbf{x}_k, \mathbf{v}_k, \ldots, \mathbf{x}_N, \mathbf{v}_N).$$ (1.2.10)

The modeling approach by methods of mathematical kinetic theory essentially means deriving an evolution equation, starting from a detailed analysis of microscopic interactions, for the distribution function. In the case of the Boltzmann equation, microscopic interactions are elastic collisions between spherical particles modeled as point masses. Collisions preserve mass, momentum, and energy, while the evolution equation is obtained by a mass conservation equation in each elementary volume of the phase space. Collision interactions are localized in space, namely, when two particles are very close each other. Only pair interactions are assumed to be significant.

It is well known that the derivation requires assumptions which cannot be fully justified from a mathematical viewpoint. Indeed, the one-particle distribution function is obtained as the marginal density of the N-particle distribution function. A rigorous derivation of the evolution equation leads to a hierarchy of equations, the BBGKY hierarchy, see Cercignani, Illner, and Pulvirenti (1994), involving all distributions from the first to the last. The hierarchy is such that the first equation, which describes the evolution on the one-particle distribution function, also involves the two-particle distribution function. The second equation, which describes the evolution on the two-particle distribution function, also involves the three-particle distribution function; and so on to the nth equation describing the distribution n-particle function, but involving the $(n+1)$-distribution.

Therefore, an evolution equation just for f_1 may only be an approximation, however useful, of physical reality. In particular, its phenomenological derivation requires the assumption of factorization of the two-particle distribution function

$$f_2 = f_2(\mathbf{x}_1, \mathbf{v}_1, \mathbf{x}_2, \mathbf{v}_2) = f_1(\mathbf{x}_1, \mathbf{v}_1) f_1(\mathbf{x}_2, \mathbf{v}_2).$$ (1.2.11)

The derivation of the Boltzmann equation is obtained within the above approximation. In particular, referring to the classic literature, the derivation consists in equating the time derivative of f in a reference volume $d\mathbf{x}\, d\mathbf{v}$ to the difference between the **gain** and **loss** terms of the particles which, due to the collisions, enter into the said volume and leave it. This concept is written as

$$\frac{df}{dt}(t, \mathbf{x}, \mathbf{v}) d\mathbf{x}\, d\mathbf{v} = (G[f, f] - L[f, f])\,(t, \mathbf{x}, \mathbf{v}) d\mathbf{x}\, d\mathbf{v}.$$ (1.2.12)

The loss term $L[f, f]$ is evaluated by computing all collisions experienced by the test particle with velocity $\mathbf{v}$ in its encounters with the field particle with velocity $\mathbf{w}$. Performing some technical calculations yields

$$L[f, f](t, \mathbf{x}, \mathbf{v}) = \int_{\mathbf{D}} |\mathbf{w} - \mathbf{v}| f(t, \mathbf{x}, \mathbf{v}) f(t, \mathbf{x}, \mathbf{w}) \, b \, d\epsilon \, db \, d\mathbf{w} \,, \qquad (1.2.13)$$

where $\mathbf{D} = \mathbb{R}^3 \times 2\pi \times [0, R]$, while ϵ and b are parameters, angular and linear, of the collision.

Analogous calculations can be performed for the gain term $G[f, f]$ related to the collisions of particles which fall into the state of the test particle:

$$G[f, f](t, \mathbf{x}, \mathbf{v}) = \int_{\mathbf{D}} |\mathbf{w}' - \mathbf{v}'| f(t, \mathbf{x}, \mathbf{v}') f(t, \mathbf{x}, \mathbf{w}') \, b' \, d\epsilon' \, db' \, d\mathbf{w}'. \quad (1.2.14)$$

Exploiting Liouville's theorem of conservation of phase volume:

$$d\mathbf{v}' \, d\mathbf{w}' = d\mathbf{v} \, d\mathbf{w} \,,$$

and suitable symmetry properties of the collision dynamics,

$$\epsilon = \epsilon', \quad b = b' \quad \text{and} \quad |\mathbf{w} - \mathbf{v}| = |\mathbf{w}' - \mathbf{v}'| \,,$$

yields the ***Boltzmann equation*** which is an evolution equation for the one-particle distribution function. This model, in the absence of an external force field, can be written as follows:

$$\left(\frac{\partial}{\partial t} + \mathbf{v} \cdot \nabla_{\mathbf{x}} \right) f = J[f, f] = G[f, f] - L[f, f] \,, \qquad (1.2.15)$$

where $f, J, G,$ and L depend on $(t, \mathbf{x}, \mathbf{v})$, with

$$G[f, f](t, \mathbf{x}, \mathbf{v}) = \int_{\mathbb{R}^3 \times \mathbb{S}^2_+} B(\mathbf{n}, \mathbf{q}) f(t, \mathbf{x}, \mathbf{v}') f(t, \mathbf{x}, \mathbf{w}') \, d\mathbf{n} \, d\mathbf{w} \,, \qquad (1.2.16)$$

and

$$L[f, f](t, \mathbf{x}, \mathbf{v}) = f(t, \mathbf{x}, \mathbf{v}) \int_{\mathbb{R}^3 \times \mathbb{S}^2_+} B(\mathbf{n}, \mathbf{q}) f(t, \mathbf{x}, \mathbf{w}) \, d\mathbf{n} \, d\mathbf{w} \,. \qquad (1.2.17)$$

The following notation has been used.

- **n** is the unit vector in the direction of the apse-line bisecting the angle between velocities $\mathbf{q} = \mathbf{w} - \mathbf{v}$ and $\mathbf{q}' = \mathbf{w}' - \mathbf{v}'$.
- $\mathbb{S}_+^2$ is the integration domain of **n**

$$\mathbb{S}_+^2 = \{ \mathbf{n} \in \mathbb{R}^3 : \quad |\mathbf{n}| = 1, \quad \mathbf{n} \cdot \mathbf{q} \geq 0 \}. \tag{1.2.18}$$

- $B(\mathbf{n}, \mathbf{q})$ is a collision kernel which depends upon the interaction potential. Detailed expressions of B are given in the pertinent literature, e.g., Cercignani, Illner, and Pulvirenti (1993).
- $\mathbf{v}, \mathbf{w}$ are the pre-collision velocities of the test and field particles, respectively, and $\mathbf{v}', \mathbf{w}'$ are the post-collision velocities, related to $\mathbf{v}$ and $\mathbf{w}$ by the relations

$$\mathbf{v}' = \mathbf{v} + \mathbf{n}\,(\mathbf{n} \cdot \mathbf{q}), \qquad \mathbf{w}' = \mathbf{w} - \mathbf{n}\,(\mathbf{n} \cdot \mathbf{q}). \tag{1.2.19}$$

Solving the mathematical problems related to the Boltzmann equation provides the distribution function and consequently the macroscopic quantities. We also recall that the equation $J(f, f) = 0$ is a functional equation which admits the **Maxwellian equilibrium solutions** which can be written as

$$\omega(t, \mathbf{x}, \mathbf{v}) = \frac{\rho(t, \mathbf{x})}{(2\pi(k/m)\Theta(t, \mathbf{x}))^{3/2}} \exp \left\{ -\frac{[\mathbf{v} - \mathbf{U}(t, \mathbf{x})]^2}{2(k/m)\Theta(t, \mathbf{x})} \right\}, \tag{1.2.20}$$

where Θ is the temperature related, at equilibrium, to the mean energy (1.2.8).

Moreover, the *H-Boltzmann functional*

$$H(t) = \int_{\mathbb{R}^3 \times \mathbb{R}^3} f \log f \, d\mathbf{x} \, d\mathbf{v} \tag{1.2.21}$$

is monotone decreasing in time towards the stable equilibrium configuration.

The above stability results have inspired simple phenomenological models of the type

$$\left(\frac{\partial}{\partial t} + \mathbf{v} \cdot \nabla_{\mathbf{x}} \right) f = J_p[f, f] = \nu[f] \big(\omega(\rho, \mathbf{U}, \mathcal{E}) - f \big), \tag{1.2.22}$$

where ν is related to the collision frequency and ω is the Maxwellian distribution function identified by the local macroscopic variables. This model is known as a **BGK-type model**, where the collision dynamics is somehow hidden in the operator which expresses the trend to equilibrium.

Particles, in the absence of an external force field, follow straight lines between two successive collisions. However, when an external field is applied, the trajectory is determined by the laws of classical mechanics, and the equation is as follows:

$$\left(\frac{\partial}{\partial t} + \mathbf{v} \cdot \nabla_{\mathbf{x}} + \mathbf{F} \cdot \nabla_{\mathbf{v}}\right) f = J[f, f] = G[f, f] - L[f, f], \qquad (1.2.23)$$

where $\mathbf{F}$ is the external positional force field acting on each of the identical particles, and the collision operator has been defined above.

The Boltzmann equation is a model derived under the assumption that the distribution function is modified only by external actions and short range interactions. On the other hand, various physical systems also involve significant long range interactions. This is the reasoning which leads to the collisionless *Vlasov equation*.

Again, the modeling refers to the test and field particles, however interactions are of long range type because field particles apply to the test particle an action which depends on their distance. The additional assumption, that each action is not influenced by the presence of the other particles, allows us to compute the overall action of the field particles over the test one. Consequently, the transport term is computed and mass conservation yields the evolution equation. In this case, the definition of mean field interactions is used.

Let us then consider the vector action $\mathcal{P} = \mathcal{P}(\mathbf{x}, \mathbf{v}, \mathbf{x}_*, \mathbf{v}_*)$ on the test particle with microscopic state $\mathbf{x}, \mathbf{v}$ due to the field particle with state $\mathbf{x}_*, \mathbf{v}_*$. The resultant action is

$$\mathcal{F}[f](t, \mathbf{x}, \mathbf{v}) = \int_{\mathbb{R}^3 \times \mathcal{D}_\Omega} \mathcal{P}(\mathbf{x}, \mathbf{v}, \mathbf{x}_*, \mathbf{v}_*)\, f(t, \mathbf{x}_*, \mathbf{v}_*)\, d\mathbf{x}_*\, d\mathbf{v}_*, \qquad (1.2.24)$$

where $\mathcal{D}_\Omega$ is the domain around the test particle where the action of the field particle is effectively felt. In other words, the action $\mathcal{P}$ decays with the distance between test and field particles and is equal to zero on the boundary of $\mathcal{D}_\Omega$. Based on these assumptions, the mean field equation is written

$$\frac{\partial f}{\partial t} + \mathbf{v} \cdot \nabla_{\mathbf{x}} f + \mathbf{F} \cdot \nabla_{\mathbf{v}} f + \nabla_{\mathbf{v}} \cdot (\mathcal{F}[f]f) = 0, \qquad (1.2.25)$$

where $\mathbf{F}$ is the positional action applied by the outer environment.

Mathematical problems related to models of the kinetic theory can be classified, as usual, into initial, initial-boundary, and boundary value problems. However the statement of the problems is somewhat different from that for models of continuous mechanics. A brief introduction is given below.

Let us first consider the initial value problem for the Boltzmann and Vlasov equations in the whole space $\mathbb{R}^3$. The problem is stated linking Eq. (1.2.15) or (1.2.25) with the initial conditions

$$f^0(\mathbf{x}, \mathbf{v}) = f(0, \mathbf{x}, \mathbf{v}) : \mathbb{R}^3 \times \mathbb{R}^3 \to \mathbb{R}_+ , \qquad (1.2.26)$$

which are assumed to decay at infinity in space.

If the characteristic lines can be identified, the problem can be written in a suitable integral (mild) form. For instance, the integral form of the initial value problem for the Boltzmann equation, when $\mathbf{F} = \mathbf{0}$, reads

$$f^\#(t, \mathbf{x}, \mathbf{v}) = f^0(\mathbf{x}, \mathbf{v}) + \int_0^t J^\#(s, \mathbf{x}, \mathbf{v}) \, ds , \qquad (1.2.27)$$

where

$$f^\#(t, \mathbf{x}, \mathbf{v}) = f(t, \mathbf{x} + \mathbf{v}\,t, \mathbf{v}) ,$$

and

$$J^\#(t, \mathbf{x}, \mathbf{v}) = J[f, f](t, \mathbf{x} + \mathbf{v}\,t, \mathbf{v}) .$$

The statement of the initial-boundary value problem requires the modeling of gas-surface interaction phenomena. In particular, two specific problems can be stated:

- The *interior domain problem*, which corresponds to a gas contained in a volume bounded by a solid surface;
- The *exterior domain problem*, which corresponds to a gas in the whole space $\mathbb{R}^3$ which contains an obstacle.

The surface of the solid wall is defined in both cases by Ω_w, and the normal to the surface directed towards the gas is $\vec{\nu}$. Moreover, in order to define the boundary conditions on a solid wall, we need to define the partial incoming and outgoing traces f^+ and f^- on the boundary Ω_w, which, for continuous f, can be defined as follows:

$$\begin{cases} f^+(\mathbf{x}, \mathbf{v}) = f(\mathbf{x}, \mathbf{v}) , & \mathbf{x} \in \Omega_w , \quad \mathbf{v} \cdot \vec{\nu}(\mathbf{x}) > 0 , \\ f^+(\mathbf{x}, \mathbf{v}) = 0 , & \mathbf{x} \in \Omega_w , \quad \mathbf{v} \cdot \vec{\nu}(\mathbf{x}) < 0 , \end{cases} \qquad (1.2.28)$$

and

$$\begin{cases} f^-(\mathbf{x}, \mathbf{v}) = f(\mathbf{x}, \mathbf{v}) , & \mathbf{x} \in \Omega_w , \quad \mathbf{v} \cdot \vec{\nu}(\mathbf{x}) < 0 , \\ f^-(\mathbf{x}, \mathbf{v}) = 0 , & \mathbf{x} \in \Omega_w , \quad \mathbf{v} \cdot \vec{\nu}(\mathbf{x}) > 0 . \end{cases} \qquad (1.2.29)$$

The boundary conditions are formally stated as follows:

$$f^+(t, \mathbf{x}, \mathbf{v}) = \mathcal{R} \, f^-(t, \mathbf{x}, \mathbf{v}) , \qquad (1.2.30)$$

where the operator $\mathcal{R}$, which maps the distribution function of the particles which collide with the surface into the one of particles leaving the surface, is characterized, for a broad range of physical problems, by the following properties.

i) $\mathcal{R}$ is linear, of local type with respect to $\mathbf{x}$, and is positive:

$$f^- \geq 0 \Rightarrow \mathcal{R} f^- \geq 0 \,. \tag{1.2.31}$$

ii) $\mathcal{R}$ preserves mass, i.e., the flux of the incoming particles equals the one of the particles which leave the surface.

iii) $\mathcal{R}$ preserves local equilibrium at the boundary: $\omega_w^+ = \mathcal{R}\,\omega_w^-$, where ω_w is the Maxwellian distribution at the wall temperature.

iv) $\mathcal{R}$ is dissipative, i.e., satisfies the inequality at the wall

$$\int_{\mathbb{R}^3} \langle \mathbf{v}, \mathbf{n} \rangle \left(f^- + \mathcal{R} f^- \right) \left(\log f + \frac{|\mathbf{v}|^2}{\Theta} \right) d\mathbf{v} \leq 0 \,. \tag{1.2.32}$$

The book by Cercignani, Illner, and Pulvirenti (1994) provides the description of some specific models of gas-surface interaction.

Therefore, formulation of the initial-boundary value problem, for the ***interior domain problem***, consists in linking the evolution equations to the initial conditions (1.2.26), and to the boundary conditions (1.2.30) on the wall Ω_w. For the ***exterior domain problem***, in addition to the boundary conditions on the wall, suitable Maxwellian equilibrium conditions are assumed at infinity. If one refers to the boundary value problem, the statement of the boundary conditions must be linked to the ***steady Boltzmann equation***, and the definitions of solutions are analogous to that we have seen for the initial value problem.

The preceding mathematical statements refer to the Boltzmann equation in the absence of an external field, and hence the trajectories of the particles are straight lines. When an external field is acting on the particles, trajectories may be explicitly determined only by solving the equations of Newtonian dynamics.

Several books, for instance Glassey (1995), provide the interested reader with a qualitative analysis of mathematical problems related to the application of the Boltzmann equation.

The works cited near the end of Section 1.1 report on several mathematical problems related to the above classical equations of mathematical kinetic theory; in particular, the derivation from the microscopic dynamics, and the qualitative analysis of the initial and initial-boundary value problems in unbounded and bounded domains. As a matter of fact, only parts of the above problems have been properly solved despite several valuable mathematical results reviewed in Villani's survey (2004). Most of these problems remain challenging research perspectives for applied mathematicians. When one looks at generalizations of mathematical kinetic theory to the modeling of complex systems in life sciences, the background of open problems should always be kept in mind.

1.3 Discrete Velocity Models

The classical models of the kinetic theory of gases are derived with the assumption that particles move freely in space with velocities which may attain all values in $\mathbb{R}^3$. On the other hand, a well-known class of models has been derived on the basis of the assumption that velocities can assume only a finite number of values. The discretization process attempts to reduce the complexity of the models and also develop suitable computational schemes. A concise introduction to discrete velocity models is given here in view of the derivation of models for active particles with discrete states.

This presentation is motivated by the fact that some living systems have microscopic activity that does effectively assume finite (not continuous) values. In this case the discretization is not regarded as a way to reduce complexity, but as a consistent approach to the modeling. What is called the ***discrete Boltzmann equation*** is a useful background for the derivation of models with discrete activity states.

The discrete models of the Boltzmann equation can be obtained assuming that particles are allowed to move with a finite number of velocities. The model is an evolution equation for the number densities $N_i = N_i(t, \mathbf{x})$ linked to the admissible velocities

$$\mathbf{v}_i, \quad \text{for} \quad i \in \mathbf{L} = \{1, \ldots, n\}. \tag{1.3.1}$$

The set $N = \{N_i\}_{i=1}^n$ corresponds, for certain aspects, to the one-particle distribution function of the continuous Boltzmann equation. The mathematical discrete kinetic theory was systematically developed in the lecture notes by Gatignol (1975), which provide a detailed analysis of the relevant aspects of the discrete kinetic theory: modeling, analysis of thermodynamic equilibrium, and application to fluid dynamics problems. The analysis mainly refers to a simple monatomic gas and to the related thermodynamic aspects. After this fundamental contribution, several further developments deal with more general physical systems: gas mixtures, chemically reacting gases, particles undergoing multiple collisions, and so on.

Various mathematical aspects, namely the qualitative analysis of the initial value and initial-boundary value problems, have been the object of continuous interest of applied mathematicians. The existing literature is reported in the review papers by Platkowski and Illner (1985) and by Bellomo and Gustafsson (1991).

The formal expression of the evolution equation is as follows:

$$\left(\frac{\partial}{\partial t} + \mathbf{v}_i \cdot \nabla_\mathbf{x} \right) N_i = J_i[N] \,, \qquad (1.3.2)$$

where

$$N_i = N_i(t, \mathbf{x}) : \quad (t, \mathbf{x}) \in [0, T] \times D_\mathbf{x} \to \mathbb{R}_+ \,, \quad i = 1, \ldots, n \,, \qquad (1.3.3)$$

where t and $\mathbf{x} \in D_\mathbf{x}$ are the time and the space variables, whereas $J_i[N]$ denote the binary collision terms

$$J_i[N] = \frac{1}{2} \sum_{j,h,k=1}^{n} A_{ij}^{hk} (N_h N_k - N_i N_j) \,. \qquad (1.3.4)$$

The terms A_{ij}^{hk} are the ***transition rates*** referred to the binary collisions

$$(\mathbf{v}_i, \mathbf{v}_j) \longleftrightarrow (\mathbf{v}_h, \mathbf{v}_k) \,, \qquad i, j, h, k \in \mathbf{L} \,, \qquad (1.3.5)$$

where the collision scheme must be such that mass, momentum, and energy are preserved.

The transition rates are positive constants which, according to the indistinguishability property of the gas particles and to the reversibility of the collisions, satisfy the following relations:

$$A_{hk}^{ij} = A_{ij}^{hk} = A_{ji}^{hk} = A_{ij}^{kh} = A_{ji}^{kh} \,. \qquad (1.3.6)$$

As for the Boltzmann equation the qualitative analysis of discrete models requires the identification of the space of collision invariants and of the ***Maxwellian state***, which is identified by the condition $J_i[N] = 0$, for $i \in \mathbf{L} = \{1, \ldots, n\}$. Moreover, for this model, a classical *H*-Boltzmann function defined as

$$H = \sum_{i \in \mathbf{L}} N_i \log N_i \qquad (1.3.7)$$

can be identified.

Therefore, the evolution equation for the *H*-Boltzmann equation can be derived by multiplying the discrete Boltzmann equation by $1 + \log N_i$ and taking the sum over $i \in \mathbf{L}$. It can be technically verified that the time derivative of this functional is nonpositive and that the equality holds if and only if the system is in a Maxwellian state.

Different models can be technically derived according to various discretization schemes as documented in the lecture notes by Gatignol (1975). For instance, when all velocities have the same modulus with six directions corresponding to the positive and negative directions of the coordinate axes, then technical calculations lead to what is called Broadwell's model, which has been applied to several interesting problems related to shock wave phenomena.

The velocities of Broadwell's model correspond to vectors joining the centers of a cube to its faces. However, models with one velocity *modulus* cannot describe phenomena with energy (temperature) variations. Therefore, Cabannes (1980) introduced a 14-velocity model with two velocity moduli. The additional velocities correspond to the vertex of the cube of Broadwell's model. This model has also been applied to the analysis of some interesting shock wave phenomena.

The interested reader is referred to the literature proposed in the book edited by Bellomo and Gatignol (2003), where a large variety of developments, following the above-cited pioneer papers, are reported.

1.4 Guidelines for Modeling Living Systems

Traditionally, methods of mathematical kinetic theory have been applied to model the evolution of large systems of interacting classical or quantum particles. An account and some bibliographic notes have been given in Sections 1.2 and 1.3. Because this book deals with the modeling of systems which are characterized by a somehow organized, or even intelligent, behavior, our approach is developed to deal with living, rather than inert, matter.

This section is focused on the statistical representation and modeling of large systems of interacting *living* entities, occasionally called subjects or individuals, which may be cells in large biological systems, individuals in a socially active population, vehicle drivers on roads, and so on. This section provides a concise introduction to the modeling of living systems. The topic will be properly developed in the following chapters.

The microscopic state always includes geometrical variables suitable to identify the particles' position and shape, and mechanical quantities related to their velocity. However, for living systems, the identification of the microscopic state requires additional variables specific to the particular system being modeled. For instance, a variable may be related to biological functions in the case of cell populations, or to the social state in the case of dynamics of populations.

Motivations to use a statistical representation of **living** systems can be found in various fields of applied sciences. Biology is a science which requires this approach, as documented in the interesting paper by Hartwell, Hopfield, Leibner, and Murray (1999). The following sentence is particularly significant:

Although living systems obey the laws of physics and chemistry, the notion of function or purpose differentiates biology from other natural sciences.

Moreover, in the same paper:

More important, what really distinguishes biology from physics are survival and reproduction, and the concomitant notion of function.

Although these sentences are specifically referred to biological systems, their validity can clearly be extended to a large variety of complex living systems.

The description of the system by mathematical kinetic theory methods essentially means defining the microscopic state of the entities interacting within a large system, and the distribution function over this state. Some preliminary definitions are necessary because the terminology related to this type of systems has not yet been unified in the literature.

Some concepts are introduced here, and more precise definitions and related analyses will be given in Chapter 2 in connection with the actual derivation of suitable classes of evolution equations.

Let us consider a system constituted by a large number of interacting entities, which will be called **active particles**, or occasionally **agents**, and which are generally organized in different interacting populations.

• The **microscopic state** of the active particles is the variable which defines their physical state. The same variable is used for all particles, yet this variable attains different values for each particle.

• The microscopic state, which is denoted by the variable $\mathbf{w}$, is formally written as follows:

$$\mathbf{w} = \{\mathbf{z}, \mathbf{q}, \mathbf{u}\} \in D_{\mathbf{w}} = D_{\mathbf{z}} \times D_{\mathbf{q}} \times D_{\mathbf{u}}, \qquad (1.4.1)$$

where $\mathbf{z}$ is the **geometrical microscopic state**, for instance, position and orientation; $\mathbf{q}$ is the **mechanical microscopic state**, for instance, linear and angular velocities; and $\mathbf{u}$ is the **social** or **biological microscopic state**, which is called the **activity**. Moreover, $D_{\mathbf{z}}$, $D_{\mathbf{q}}$, and $D_{\mathbf{u}}$ refer to the domains of existence of these variables, while the space $D_{\mathbf{w}}$ of the microscopic states is called the **state space**.

• As a particular case, the geometrical and mechanical states of the system are identified, respectively, by position $\mathbf{x}$ and velocity $\mathbf{v}$, consequently,

$$\mathbf{w} = \{\mathbf{x},\,\mathbf{v},\,\mathbf{u}\} \in D_{\mathbf{w}} = D_{\mathbf{x}} \times D_{\mathbf{v}} \times D_{\mathbf{u}}\,. \qquad (1.4.2)$$

• The dimensions of the active particles are small with respect to the dimensions of the environment identified by the domain $\Omega \subseteq \mathbb{R}^3$ which contains them, and in some cases small even with respect to the average distance separating them.

• The overall state of the system is described by the distribution function

$$f = f(t, \mathbf{w}) = f(t, \mathbf{z}, \mathbf{q}, \mathbf{u}) : \quad [0, T] \times D_{\mathbf{w}} \to \mathbb{R}_+\,, \qquad (1.4.3)$$

which is called the ***generalized distribution function***.

• $f(t, \mathbf{w})\,d\mathbf{w}$ denotes the number of entities whose state, at time t, is in the domain $[\mathbf{w}, \mathbf{w} + d\mathbf{w}]$. When the geometrical and mechanical microscopic states simply refer to position $\mathbf{x}$ and velocity $\mathbf{v}$, the distribution function is written: $f = f(t, \mathbf{x}, \mathbf{v}, \mathbf{u})$.

• Pair interactions refer to the ***test*** or ***candidate*** active particle interacting with a ***field*** active particle. These encounters modify the microscopic state of the particles and may also cause generation or destruction of particles.

• The test particle is selected as representative of the whole system. Hence, the distribution function refers to this particle.

If f is known, then macroscopic gross variables can be computed, under suitable integrability properties, as moments weighted by the above distribution function. The difference with respect to the classical distribution function is that the calculation of mechanical variables requires the additional integration of the activity variable. For instance, the zeroth-order moment, in the simple case of mechanical microscopic state identified by position and velocity, gives the density of active particles:

$$n = n(t, \mathbf{x}) = \int_{D_{\mathbf{v}} \times D_{\mathbf{u}}} f(t, \mathbf{x}, \mathbf{v}, \mathbf{u})\, d\mathbf{v}\, d\mathbf{u}\,, \qquad (1.4.4)$$

while the number of particles is obtained by an additional integration over the space variable:

$$N = N(t) = \int_{D_{\mathbf{x}} \times D_{\mathbf{v}} \times D_{\mathbf{u}}} f(t, \mathbf{x}, \mathbf{v}, \mathbf{u})\, d\mathbf{x}\, d\mathbf{v}\, d\mathbf{u}\,. \qquad (1.4.5)$$

Integration over the mechanical variables, provides, as we shall see in Chapter 2, macroscopic information on the macroscopic action of the activity variable.

Position and velocity are not always enough to identify the microscopic mechanical state. For instance, in mathematical biology, cells are the active particles of multicellular systems with a shape which may play a relevant role in microscopic interactions, and in the modeling of vehicles on a road, the vehicle size plays a nonnegligible role. Therefore, additional variables and calculations are necessary.

A mathematical model of the kinetic theory for active particles is an evolution equation for the preceding distribution function f over the microscopic state of large systems of interacting particles. Mathematical problems are obtained by linking the evolution equation to suitable initial and boundary conditions. The solution of problems gives the distribution function; subsequently, weighted moments allow us to compute macroscopic quantities.

The derivation of the evolution equation requires us to model microscopic interactions that, as in classical kinetic theory, can be of short or long range type between field and test individuals. Subsequently, suitable conservation equations in the elementary volume of the state space lead to the derivation of models. We shall see in Chapter 2 how to transfer these introductory reasonings into suitable mathematical terms.

We anticipate that the mathematical structure is analogous to that of models for classical particles with the addition of suitable proliferative/destructive terms. For instance, for models with short range interactions the formal structure of the equation is as follows:

$$\left(\frac{\partial}{\partial t} + \mathbf{v} \cdot \nabla_{\mathbf{x}} + \mathbf{F} \cdot \nabla_{\mathbf{v}} \right) f = J[f, f] + S_s[f, f] , \qquad (1.4.6)$$

where $J[f, f]$ corresponds to interactions which modify the microscopic state without modifying the number of active particles, and $S_s[f, f]$ corresponds to proliferation or destruction of active particles due to short range interaction.

Similarly, the structure of the equation for long range interactions is as follows:

$$\frac{\partial f}{\partial t} + (\mathbf{v} \cdot \nabla_{\mathbf{x}})f + \mathbf{F} \cdot \nabla_{\mathbf{v}} f + \nabla_{\mathbf{w}} \cdot \mathcal{F}[f] = S_m[f, f] , \qquad (1.4.7)$$

where $S_m[f, f]$ corresponds to proliferation or destruction of active particles due to mean field interaction.

The main difference is that the derivation of suitable expressions of the terms $J[f, f]$, $S_s[f, f]$, $S_m[f, f]$ and the vector $\mathcal{F}[f]$ requires a detailed analysis of microscopic interactions, which are stochastic in the case of living systems.

Interactions are modeled by a stochastic game theory because the microscopic activity modifies rules of classical mechanics so that the background framework of Newtonian mechanics, valid for systems of classical particles, does not hold anymore. Therefore, alternative paradigms should be found in modeling microscopic interactions, where mechanical variables may influence interactions at the level of the activity and vice versa. This is a crucial aspect.

Some real systems of active particles are such that their activity is a discrete, rather than continuous, variable identified by a set of elements $I_k = \{\mathbf{u}_k\}$. In this case, the distribution function is a set where each element corresponds to each element of I_k. The reference model of the mathematical kinetic theory is the discrete, rather than the continuous, Boltzmann equation. Various applications will be developed in the second part of this book.

A very simple example of a model of stochastic kinetic theory is reported here to introduce the contents of the forthcoming chapter. This example refers to a population dynamics model with kinetic interaction, proposed by Jager and Segel (1992) for studying the evolution of a physical state, called **dominance**, that characterizes the ability of some individuals to dominate the others in certain populations of insects. The biological system was experimentally observed by Hogeweg and Hesper (1983).

The model consists of a nonlinear integro-differential equation, that defines the evolution of the probability density function over the dominance. Individuals can be found in a state described by a variable $u \in [0, 1]$, which identifies the **microscopic state**, and which should be regarded as a dimensionless, normalized, real-valued independent variable. The description of the overall state of the system is given by the probability density function

$$f = f(t, u) \, : \, [0, T] \times [0, 1] \to \mathbb{R}_+ \, . \tag{1.4.8}$$

Therefore, the probability of finding, at time t, an individual in the state interval $[u_1, u_2]$ is given by

$$\mathcal{P}(t, u \in [u_1, u_2]) = \int_{u_1}^{u_2} f(t, u) \, du \, . \tag{1.4.9}$$

The mathematical model is derived assuming that the microscopic state is modified by encounters and that only binary encounters play a role in the game. Moreover, the derivation requires a detailed analysis of microscopic interactions, which can be identified by the following quantities:

The **encounter rate** between pairs of individuals in the states u and v, which is identified by the term $\eta(u, v) \geq 0$.

The **probability density** that an individual in the state v ends up in the state u conditionally to an encounter with an individual in the state w; it has a density with respect to the variable u denoted by $\varphi(v, w; u) \geq 0$.

If no external action modifies the internal state, the expression of the mathematical model follows:

$$\frac{\partial f}{\partial t}(t, u) = \int_0^1 \int_0^1 \eta(v, w)\varphi(v, w; u)f(t, v)f(t, w)\, dv\, dw$$

$$- f(t, u) \int_0^1 \eta(u, v)f(t, v)\, dv\,. \qquad (1.4.10)$$

This model has been derived simply by equating, in the elementary volume $[u, u + du]$, the rate of increase of the number of individuals with state u to the net flow (inflow minus outflow) of individuals due to interactions. In particular, the assumption that φ is a probability density means that the number of individuals of each population is preserved.

This section has simply outlined the formal structure of the models. A detailed analysis will be developed in Chapters 2–4 by the derivation of a general mathematical framework which includes, as special cases, the classical models introduced in Section 1.2.

1.5 Purpose and Plan of the Book

This book studies the modeling of large complex systems in life sciences by suitable developments of the methods of mathematical kinetic theory. These systems are composed of a large number of **active particles** or **agents** with interactions which modify their microscopic state, which includes the **activity**, and may even generate proliferative and/or destructive events.

The first part of the book concerns some methodological aspects, and specifically the derivation of general evolution equations suitable to be specialized to model some particular systems of interest in various field of applied sciences. The specialization consists in a detailed modeling of microscopic interactions specific to the system to be modeled. The second part deals with various modeling issues and applications. The presentation follows a common style for all chapters:

i) Phenomenological interpretation of the physical system in view of the mathematical modeling and identification of a microscopic variable suitable for describing the state of each individual;

ii) Derivation of the mathematical model by mathematical kinetic theory methods for active particles. Models are obtained by a balance of particles in each elementary volume of the space of microscopic states;

iii) Simulations, parameter sensitivity analysis, and critical analysis of the model;

iv) An overview of ideas for improving the existing models with special attention to applications.

Mathematical problems related to the application of models to the description of real world phenomena often involve interesting and challenging analytic problems. This topic, which represents an interesting research perspective for applied mathematicians, is constantly posed to the reader by means of bibliographical indications and motivations to tackle several open problems. However, greater emphasis is given to modeling and simulations.

The contents of the remaining chapters are as follows.

• Chapter 2 derives the evolution equations for the one-particle distribution function over the microscopic state of systems involving a large number of interacting active particles. Two types of equations are derived corresponding, respectively, to short and long range interactions. These equations are regarded as a general mathematical framework which may include, as special cases, various models of living systems, as well as the classical models of kinetic theory for inert particles. In particular, it can be shown how the equations proposed in this chapter include, under suitable technical assumptions, the classical models of kinetic theory: the Boltzmann and the Vlasov equations.

• Chapter 3 further develops these topics to deal with some technical generalizations that refer to modeling mixtures with mixed-type long and short range interactions and to proliferative events that generate active particles in a population different from those of the interacting pairs. This chapter looks at research perspectives, while the previous one contains the fundamental structures that are used in the modeling of Chapters 5–8.

• Chapter 4 focuses on the modeling of systems for which the microscopic state attains a finite number of states rather than a continuous one. The microscopic state of some models of interest in applied sciences is a finite-dimensional variable rather than a continuous variable. Therefore, this class of equations technically refers to the discrete Boltzmann equation and is proposed for modeling purposes rather than with the aim of reducing computational complexity. Various applications presented in the book will refer to this specific mathematical framework with discrete states.

The various classes of equations in Chapters 2–4 are proposed in view of the applications we study in the second part of the book, which refers to the design and analysis of some classes of models of interest in applied sciences. In some cases, different frameworks can be used to model the same system. A precise rule cannot be given, as the choice of method depends on the strategy developed by the model designer.

The book is not limited to what is already available in the literature; various perspective ideas for the development of new models are offered to the attention of the reader. The presentation mainly refers to modeling aspects, with some simple simulations developed to visualize the predictive ability of the models.

• Chapter 5 discusses the modeling of various phenomena of social competition related to individuals characterized by a microscopic state corresponding to their social collocation. Microscopic interactions modify such a state, while the overall system is described by a probability density distribution over the microscopic state. Models are analyzed with special attention to their predictive ability. This means that the investigation is addressed to analyze how certain microscopic interactions may generate different types of evolutions. Some technical generalizations concerning systems of several interacting populations refer to systems where the total number of interacting particles is still constant in time, but may change in time within each population.

This chapter refers to models in spatial homogeneity. Interactions modify the activity, whereas the space dynamics does not play a relevant role. Some perspective ideas to introduce the modeling of a space dynamics are proposed in view of the chapters which follow, where this aspect is studied. Moreover, this chapter essentially refers to models with discrete states. However, suitable indications to derive continuous models are brought to the attention of the reader.

• Chapter 6 deals with vehicular traffic flow modeling. The car-driver system is considered as an active particle. Microscopic interactions generate slowing down or acceleration of vehicles which may pass (or be passed by) other vehicles. The space variable is an essential feature of the model and also plays a relevant role in the mathematical description of microscopic interactions. This aspect generates modeling and computational problems that require careful analysis.

As we shall see, the existing literature proposes a modeling approach essentially based on mechanical ideas typical of inert matter. However, we after some perspective ideas towards the development of tools which we hope are consistent with the requirements of modeling living matter.

The class of models in Chapters 5 and 6 is such that the number of active particles is constant in time, and in some cases, even homogeneously distributed in space. These features definitely reduce the complexity problems that need to be tackled for biological systems where interactions are accompanied by reproductive and destructive events.

• Chapter 7 is focused on models of the immune competition between cells of an aggressive invasive guest and cells of the immune system. Again the derivation of models must deal with the analysis of microscopic interactions, which now may generate proliferation and/or destruction of cells. There is a substantial difference between these models and the class of models of the preceding chapters due to the lack of conservation of mass. The class of models in this chapter refers to a recent book by Bellouquid and Delitala (2006).

Moreover, some new developments and perspectives related to models where the number of populations is variable in time are introduced. The possibility that interactions between particles belonging to two different populations may generate a daughter cell in a different, maybe new, population is properly considered. This approach may be used to model the onset of pathological states and their genetic degeneration.

• Chapter 8 deals with the modeling of crowds, where the dynamics of active particles are subject to mutual attractive or repulsive forces. Therefore, a complex dynamics is generated, that may be randomly perturbed in conditions of panic. If this class of models refers to closed domains without inlet and outlet flows, the number of particles is constant in time, otherwise the number of particles evolves in time.

Several complexity problems are generated by the difficulty in modeling microscopic interactions, where the interplay between mechanics and individual organized behavior plays a relevant role.

The modeling also considers situations of panic, which can substantially modify the evolution of the system. The essential difference of this class of models with respect to those of Chapter 6 is that drivers have similar strategies, which are not substantially modified by environmental conditions. However, in crowds, the strategy changes if the environmental conditions are modified. Moreover, some perspective ideas are provided to generalize the modeling approach to crowds to the more complex case of swarms, where the domain containing the active particles evolves in time.

• Chapter 9 goes back both to modeling perspectives, and to methodological aspects along speculations induced by a critical analysis of the various applications in the book concerning both modeling topics and analytic problems.

Chapter 9 first provides an overview of a variety of models which refer to the mathematical frameworks of Chapters 2–4. The presentation is concise so that the interested reader can develop a careful modeling process following the methods offered in Chapters 5–8. Subsequently, a critical analysis of the preceding chapters focused on the mathematical difficulties in dealing with living systems is developed, and suggestions for future research perspectives concerning modeling complex living systems are proposed.

Finally, this chapter analyzes some methodological aspects concerning multiscale modeling focused on systems where macroscopic and microscopic models must be properly matched. For instance, sometimes the modeling needs macroscopic models in some space domains, while in confining domains kinetic type equations must be used.

The selection of the classes of models reviewed in Chapters 5–8 is the personal choice of the author. Different ones may be proposed, as discussed in the last chapter. Applications have been selected with the criterion of increasing the complexity of the mathematical structures needed for the design of models. For instance, models of social dynamics refer to spatially homogeneous systems, whereas the traffic flow models of Chapter 6 need a space structure which generates various complexity issues in the modeling microscopic interactions due to the presence of both mechanical variables and activity. Chapter 7 refers to models with a variable number of active particles due to the presence of proliferative and/or destructive events, and also a variable number of equations due to the onset of daughter particles in a population different from that of the interacting pair. Chapter 8 analyzes a complex interaction dynamics in space between active particles. Interaction rules may change consistently with changes in the environment. The common feature of all these mathematical models is the attempt to describe the collective behavior starting from the microscopic dynamics.

A common line is used for the presentation of the various models. First a detailed description of the models available in the existing literature is reported; then various suggestions are given to design new models consistent with physical reality. These perspectives are developed exploiting the entire variety of mathematical frameworks offered in Chapters 2–4. These ideas can be regarded as suggestions for future research activity that refer both to modeling and analytic problems related to the applications of models.

Finally, let us mention again that the ***aim of this book*** is to develop mathematical methods and tools, for the modeling of living systems. The ***background idea*** is that models of living systems require mathematical tools—maybe technically far more complex—substantially different from those used to model inert matter.

Some interesting considerations are offered by Robert May (2004):

In the physical sciences, mathematical theory and experimental investigation have always marched together. Mathematics has been less intrusive in the life sciences because they have been largely descriptive lacking the invariance principles and fundamental constants of physics.

Mathematics contribute to this challenging objective. Going back to Charles Darwin, this effort is encouraged:

I have deeply regretted that I did not proceed far enough at least to understanding something of the great leading principles of mathematics; for men thus endowed seem to have an extra sense.

Motivations to attempt the description of living systems by mathematical equations are undoubtedly significant, and this book is offered as a first step towards the above outlined ambitious objective.

It may be that the mathematical tools available in the literature are not yet sufficient for these purposes. Yet applied mathematicians are fatally attracted to the search for new methods and mathematical structures focused on the description of the collective behavior starting from microscopic dynamics.

The analysis of models generates a variety of analytic problems which are sufficiently complex to generate a powerful attraction for the intellectual energies of applied mathematicians. This can be an additional motivation to study the modeling and analysis of living systems.

2

Mathematical Structures of the Kinetic Theory for Active Particles

---

2.1 Introduction

The microscopic state of the active particles of large living systems includes, in addition to geometric and mechanical variables, a set of variables, called the activity, suitable to describe their organized behavior. For instance activity variables can refer to their biological and social state.

Modeling complex living systems characterized by these specific features requires mathematical methods suitable for capturing the relevant aspects of the phenomenology of the systems. This chapter focuses on this topic and specifically derives suitable evolution equations that constitute a general mathematical framework to be specialized towards modeling specific systems in life sciences.

This framework consists of a system of integro-differential equations which defines the evolution of the probability distributions over the microscopic state of large systems of interacting active particles. The derivation of the evolution equation is obtained, by suitable developments of the methods of mathematical kinetic theory, according to the following guidelines already introduced in Chapter 1:

i) Assessment of the microscopic state of the active particles and of the probability distribution function over that state;

ii) Modeling of microscopic interactions which may be localized in space or long range;

iii) Derivation of an evolution equation for the above-mentioned distribution function by means of a balance equation, where the variation rate of the number of particles in the elementary volume of the state space is equated to the net flux, into such a volume, generated by interactions or transport.

The kinetic theory developed in this chapter is proposed within a self-contained presentation, so that the reader can follow the guidelines even if he/she is not fully familiar with the methods of classical kinetic theory. The mathematical structures presented in this chapter cannot yet be considered models. As we shall see, specific models are obtained only after the modeling of interactions at the microscopic scale.

The contents are presented through the following five addition sections.

– Section 2.2 introduces the the concepts of the microscopic state of active particles and the distribution function. Subsequently, it is shown how the distribution function can be used to compute macroscopic information on the overall state of the system by weighted moments. First a system consisting of one population is presented, subsequently technical generalizations to systems with several interacting populations are developed.

– Section 2.3 derives a mathematical framework for the modeling of microscopic interactions between active particles. The modeling takes into account the reciprocal influence between mechanical and activity variables by interaction schemes which may be localized or may involve long range dynamics. The activity variable can modify the mechanical interactions, which otherwise should follow the laws of classical mechanics.

– Section 2.4 shows how the modeling of microscopic interactions leads to the derivation of the evolution equations for the distribution function that define the mathematical framework offered by kinetic theory for active particles. These equations include, under particular assumptions on the microscopic states and interactions, the classical Boltzmann and Vlasov equations.

– Section 2.5 discusses some technical particularizations which can be useful for the applications. Specifically, it is shown how the mathematical structures can be particularized to model systems with dominant mechanical interactions, where the distribution over the activity remains constant in time. Analogously, equations for dominant activity interactions can be derived, where the distribution over the microscopic mechanical variables remains constant in time.

– Section 2.6 provides a critical analysis on the use of the mathematical methods developed in this chapter in view of the derivation of specific models and looks at further development of the mathematical tools.

2.2 The Generalized Distribution Function

Let us consider a physical system constituted by a large number of active particles. The physical microscopic state of the particles is identified by a variable suitable to describe their state, while the overall state of the whole system is described by a probability distribution over the microscopic state of the particles. This section provides some detailed definitions to formalize these concepts; some of them were introduced in Section 1.4. Subsequently, it is shown how suitable weighted moments can be computed to obtain the information on the macroscopic behavior of the system.

Some notation must be stated to define precisely the microscopic state of each active particle and the statistical description of the system. Consider first the relatively simple case of active particles whose geometrical state is a point of a Euclidean space:

Definition 2.2.1. *The physical variable used to describe the state of each active particle is called the **microscopic state**, which is denoted by the variable **w** formally written as follows:*

$$\mathbf{w} = \{\mathbf{x}, \mathbf{v}, \mathbf{u}\} \in D_{\mathbf{w}} = D_{\mathbf{x}} \times D_{\mathbf{v}} \times D_{\mathbf{u}}, \qquad (2.2.1)$$

where **x**, *called the **geometrical microscopic state**, identifies the position,* **v**, *called the **mechanical microscopic state**, is the velocity, and* **u** *is the **activity**, which may have a different meaning for each particular system. The space of the microscopic states is called the **state space**.*

Definition 2.2.2. *The description of the overall state of the system is defined by the distribution function*

$$f = f(t, \mathbf{w}) = f(t, \mathbf{x}, \mathbf{v}, \mathbf{u}), \qquad (2.2.2)$$

*which is called the **generalized distribution function**, such that* $f(t, \mathbf{w})$ *d**w** denotes the number of active particles whose state, at time* t, *is in the elementary volume* $[\mathbf{w}, \mathbf{w} + d\mathbf{w}]$. *Under suitable integrability assumptions, the number of particles in the domain* $\Lambda \subset D_{\mathbf{x}}$ *is given by*

$$n_{\Lambda}(t) = \int_{\Lambda \times D_{\mathbf{v}} \times D_{\mathbf{u}}} f(t, \mathbf{x}, \mathbf{v}, \mathbf{u}) \, d\mathbf{x} \, d\mathbf{v} \, d\mathbf{u}.$$

Definition 2.2.3. *Pair interactions refer to the **test** or **candidate** active particle interacting with a **field** active particle. The distribution function stated in Definition 2.2.2. refers to the test active particle.*

Remark 2.2.1. *If the number of active particles is constant in time, then the distribution function can be normalized with respect to such a number and can be regarded as a probability density. For all practical cases, it is convenient to normalize the distributions f with respect to the number, or* **size**, *of active particles at t = 0.*

Remark 2.2.2. *The above assessment of the microscopic state can be generalized by adding the* **angular configuration** *and the* **rotational velocity**, *i.e., angular velocities, respectively* **y** *and* **p**. *In this case, the distribution function is written as:*

$$f = f(t, \mathbf{x}, \mathbf{y}, \mathbf{v}, \mathbf{p}, \mathbf{u}) \,.$$

The calculations developed in what follows refer, for simplicity of notation, to the relatively simpler case of Definitions 2.2.1–2.2.2. Generalizations are technical.

If f is known, macroscopic gross variables can be computed, under suitable integrability properties, as moments weighted by the above distribution function. Specifically, marginal densities are obtained by integrating over some of the components of the variable $\mathbf{w}$. For instance, marginal densities may refer either to the generalized distribution over the mechanical state

$$f^m(t, \mathbf{x}, \mathbf{v}) = \int_{D_\mathbf{u}} f(t, \mathbf{x}, \mathbf{v}, \mathbf{u}) \, d\mathbf{u} \,, \tag{2.2.3}$$

or over the activity variable

$$f^a(t, \mathbf{u}) = \int_{D_\mathbf{x} \times D_\mathbf{v}} f(t, \mathbf{x}, \mathbf{v}, \mathbf{u}) \, d\mathbf{x} \, d\mathbf{v} \,. \tag{2.2.4}$$

Order zero moments provide information on the size. For instance, the *local size* is given by

$$n[f](t, \mathbf{x}) = \int_{D_\mathbf{v} \times D_\mathbf{u}} f(t, \mathbf{x}, \mathbf{v}, \mathbf{u}) \, d\mathbf{v} \, d\mathbf{u} = \int_{D_\mathbf{v}} f^m(t, \mathbf{x}, \mathbf{v}) \, d\mathbf{v} \,, \tag{2.2.5}$$

where square brackets are used, here and in what follows, to indicate that n can be interpreted as an operator over the distribution function f: in this specific case it is simply an integration of f over the domain $D_\mathbf{v} \times D_\mathbf{u}$.

Integration over the volume $D_\mathbf{x}$ containing the active particles gives the *total size*:

$$N[f](t) = \int_{D_\mathbf{x}} n[f](t, \mathbf{x}) \, d\mathbf{x} = \int_{D_\mathbf{u}} f^a(t, \mathbf{u}) \, d\mathbf{u} \,, \tag{2.2.6}$$

which may depend on time due to birth and death processes related to interactions, and to the flux of active particles through the boundaries of the volume.

First-order moments provide linear macroscopic quantities related both to **mechanics** and **activity**. For instance, the **mass velocity** of active particles at time t in position $\mathbf{x}$ is defined by

$$\mathbf{U}[f](t, \mathbf{x}) = \frac{1}{n[f](t, \mathbf{x})} \int_{D_{\mathbf{v}} \times D_{\mathbf{u}}} \mathbf{v} \, f(t, \mathbf{x}, \mathbf{v}, \mathbf{u}) \, d\mathbf{v} \, d\mathbf{u} \,. \qquad (2.2.7)$$

Focusing on the activity functions, linear moments can be called the **activation** at time t in position $\mathbf{x}$, related to the components u_j of the variable $\mathbf{u}$, and are computed as follows:

$$a_j = a_j[f](t, \mathbf{x}) = \int_{D_{\mathbf{v}} \times D_{\mathbf{u}}} u_j f(t, \mathbf{x}, \mathbf{v}, \mathbf{u}) \, d\mathbf{v} \, d\mathbf{u} \,, \qquad (2.2.8)$$

while the **activation density** is given by

$$\begin{aligned} a_j^d = a_j^d[f](t, \mathbf{x}) &= \frac{a_j[f](t, \mathbf{x})}{n[f](t, \mathbf{x})} \\ &= \frac{1}{n[f](t, \mathbf{x})} \int_{D_{\mathbf{v}} \times D_{\mathbf{u}}} u_j f(t, \mathbf{x}, \mathbf{v}, \mathbf{u}) \, d\mathbf{v} \, d\mathbf{u} \,. \end{aligned} \qquad (2.2.9)$$

Global quantities are obtained by integrating over the space variable. For instance, the **total activity** is given by

$$A_j = A_j[f](t) = \int_{D_{\mathbf{x}}} a_j[f](t, \mathbf{x}) \, d\mathbf{x} \,, \qquad (2.2.10)$$

while the **total activity density** is defined as follows:

$$\mathcal{A}_j = \mathcal{A}_j[f](t) = \int_{D_{\mathbf{x}}} a_j^d[f](t, \mathbf{x}) \, d\mathbf{x} \,. \qquad (2.2.11)$$

Similar calculations can be developed for higher-order moments if desired for the applications. For instance, second-order moments refer to the **translational mechanical energy**:

$$E[f](t, \mathbf{x}) = \frac{1}{n(t, \mathbf{x})} \int_{D_{\mathbf{v}} \times D_{\mathbf{u}}} [\mathbf{v} - \vec{\xi}(t, \mathbf{x})]^2 \, f(t, \mathbf{x}, \mathbf{v}, \mathbf{u}) \, d\mathbf{v} \, d\mathbf{u} \,, \qquad (2.2.12)$$

or to the **quadratic activation**:

$$e_j = e_j[f](t, \mathbf{x}) = \int_{D_\mathbf{v} \times D_\mathbf{u}} u_j^2 f(t, \mathbf{x}, \mathbf{v}, \mathbf{u}) \, d\mathbf{v} \, d\mathbf{u}, \qquad (2.2.13)$$

while the **quadratic activation density** is given by

$$\varepsilon_j = \varepsilon_j[f](t, \mathbf{x}) = \frac{e_j[f](t, \mathbf{x})}{n[f](t, \mathbf{x})}. \qquad (2.2.14)$$

The physical meaning of e_j and ε_j is respectively the energy and energy density expressed by the activity variable; the detailed meaning depends on the specific system under consideration. The energy is an important quantity as it can play a role in the strategy developed by the active particles. For instance, when the energy passes a certain threshold the strategy developed by the active particles may substantially change.

If the microscopic state includes additional variables related to space dimensions, for instance angular variables and velocity $\mathbf{y}$ and $\mathbf{p}$, analogous calculations can be developed. For instance the activity distribution is given as follows:

$$f^a(t, \mathbf{u}) = \int_{D_\mathbf{x} \times D_\mathbf{y} \times D_\mathbf{v} \times D_\mathbf{p}} f(t, \mathbf{x}, \mathbf{y}, \mathbf{v}, \mathbf{p}, \mathbf{u}) \, d\mathbf{x} \, d\mathbf{y} \, d\mathbf{v} \, d\mathbf{p}, \qquad (2.2.15)$$

while the density is given by

$$n[f](t, \mathbf{x}) = \int_{D_\mathbf{y} \times D_\mathbf{v} \times D_\mathbf{p} \times D_\mathbf{u}} f(t, \mathbf{x}, \mathbf{y}, \mathbf{v}, \mathbf{p}, \mathbf{u}) \, d\mathbf{y} \, d\mathbf{v} \, d\mathbf{p} \, d\mathbf{u}. \qquad (2.2.16)$$

All the above calculations can be straightforwardly applied to systems constituted by n interacting populations of active particles labeled by the indexes $i = 1, \ldots, n$, where each population is characterized by a different way of organizing their peculiar activities and by different interactions with the other populations. The description is now given by the one-particle distribution function

$$f_i = f_i(t, \mathbf{w}) = f_i(t, \mathbf{x}, \mathbf{v}, \mathbf{u}), \qquad (2.2.17)$$

of the ith population, while the whole set of distribution functions for all populations is denoted by $\mathbf{f} = \{f_i\}$.

Calculations are analogous to those already seen above. For instance, the **local size** of the ith population is given by

$$n_i[f_i](t, \mathbf{x}) = \int_{D_\mathbf{v} \times D_\mathbf{u}} f_i(t, \mathbf{x}, \mathbf{v}, \mathbf{u}) \, d\mathbf{v} \, d\mathbf{u}. \qquad (2.2.18)$$

The local initial size of the ith population, at $t = 0$, is denoted by n_{i0}, while the local size for all populations is denoted by n_0 and is given by

$$n_0[\mathbf{f}_0](\mathbf{x}) = \sum_{i=1}^{n} n_{i0}(\mathbf{x}), \qquad (2.2.19)$$

where $\mathbf{f}_0 = \mathbf{f}(t = 0)$ denotes the initial distribution for all populations.

Integration over the volume $D_\mathbf{x}$ containing the particles gives the **total size** of the ith population:

$$N_i(t) = \int_{D_\mathbf{x}} n_i(t, \mathbf{x}) \, d\mathbf{x}, \qquad (2.2.20)$$

which depends on time due to proliferative or destructive interactions, and to the flux of particles through the boundaries of the volume. The **total size** $N = N(t)$ of all populations is given by the sum of all N_i.

Marginal densities refer either to the generalized distribution over the mechanical state

$$f_i^m(t, \mathbf{x}, \mathbf{v}) = \int_{D_\mathbf{u}} f_i(t, \mathbf{x}, \mathbf{v}, \mathbf{u}) \, d\mathbf{u}, \qquad (2.2.21)$$

or to the generalized distribution over the microscopic activity:

$$f_i^a(t, \mathbf{u}) = \int_{D_\mathbf{x} \times D_\mathbf{v}} f_i(t, \mathbf{x}, \mathbf{v}, \mathbf{u}) \, d\mathbf{x} \, d\mathbf{v}. \qquad (2.2.22)$$

First-order moments provide either **linear mechanical macroscopic** quantities, or **linear activity macroscopic** quantities. For instance, the mass velocity of particles, at time t in position $\mathbf{x}$, is defined by

$$\mathbf{U}[f_i](t, \mathbf{x}) = \frac{1}{n_i[f_i](t, \mathbf{x})} \int_{D_\mathbf{v} \times D_\mathbf{u}} \mathbf{v} \, f_i(t, \mathbf{x}, \mathbf{v}, \mathbf{u}) \, d\mathbf{v} \, d\mathbf{u}. \qquad (2.2.23)$$

Focusing on activity terms, linear moments related to each jth component of the state $\mathbf{u}$, related to the ith population, can be called the **local activation** at time t in position $\mathbf{x}$, and are computed as follows:

$$a_{ij} = a_j[f_i](t, \mathbf{x}) = \int_{D_\mathbf{v} \times D_\mathbf{u}} u_j f_i(t, \mathbf{x}, \mathbf{v}, \mathbf{u}) \, d\mathbf{v} \, d\mathbf{u}, \qquad (2.2.24)$$

while the *local activation density* is given by

$$a_{ij}^d = a_j^d[f_i](t, \mathbf{x}) = \frac{a_j[f_i](t, \mathbf{x})}{n_i[f_i](t, \mathbf{x})}$$

$$= \frac{1}{n_i[f_i](t, \mathbf{x})} \int_{D_\mathbf{v} \times D_\mathbf{u}} u_j f_i(t, \mathbf{x}, \mathbf{v}, \mathbf{u}) \, d\mathbf{v} \, d\mathbf{u} \,. \qquad (2.2.25)$$

Global quantities are obtained by integrating over space. A different interpretation can be given for each of the quantities above. Large values of a_{ij} may be due to a large number of particles with relatively small values of the jth component of the activity, but also to a small number of particles with relatively large values of the above component, while a_{ij}^d allows us to identify the size of the mean value of the activation.

Similar calculations can be developed for higher-order moments if motivated by the application. For instance, the *local quadratic activity* can be computed as second-order moments:

$$e_{ij} = e_j[f_i](t, \mathbf{x}) = \int_{D_\mathbf{v} \times D_\mathbf{u}} u_j^2 f_i(t, \mathbf{x}, \mathbf{v}, \mathbf{u}) \, d\mathbf{v} \, d\mathbf{u} \,, \qquad (2.2.26)$$

while the *local quadratic density* is given by

$$\varepsilon_{ij} = \varepsilon_j[f_i](t, \mathbf{x}) = \frac{e_j[f_i](t, \mathbf{x})}{n_i[f_i](t, \mathbf{x})}$$

$$= \frac{1}{n_i[f_i](t, \mathbf{x})} \int_{D_\mathbf{v} \times D_\mathbf{u}} u_j^2 f_i(t, \mathbf{x}, \mathbf{v}, \mathbf{u}) \, d\mathbf{v} \, d\mathbf{u} \,. \qquad (2.2.27)$$

The above description refers to the general case when all components of the microscopic state are well mixed. However, it may happen either that only some of the components play a role in the description of the system, or that the distribution over the activity variable can be factorized with respect to the mechanical one. Specifically, the following cases can be indicated in view of the analysis of specific models.

• The *microscopic state is identified only by the activity variable*. Namely, space and velocity variables are not significant to describe the microscopic state of the active particles. In this case, the overall state of the system is described by the distribution function (2.2.4) over the activity variable only. This case has to be distinguished from that corresponding to spatial homogeneity with constant probability density distribution over the velocity variable. The description of the system is delivered by the distribution function (2.2.4), but the velocity variable no longer plays a role in the description of the system.

• The *microscopic state is identified only by the space and activity variables*. Namely, the velocity variable is not significant to describe the microscopic state. In this case, it is important to know the localization of the active particles, but not their velocity.

• The *distribution function over the activity variable is constant in time* and described by a probability density $P(\mathbf{u})$. Therefore, the overall state of the system is described by a distribution function which can be factorized as follows:

$$f = f^m(t, \mathbf{x}, \mathbf{v})P(\mathbf{u}) \,, \tag{2.2.28}$$

which, when all particles have the same state $\mathbf{u}_0$, is written as

$$f = f^m(t, \mathbf{x}, \mathbf{v})\delta(\mathbf{u} - \mathbf{u}_0) \,, \tag{2.2.29}$$

where the distribution over the activity variable is a Dirac delta function.

These two cases (2.2.28) and (2.2.29) need to be distinguished from the spatially homogeneous description with constant probability density distribution over the velocity variable. The system is still described by the distribution function (2.2.4), but the velocity variable no longer has a role in the modeling of the system.

The examples of models presented in the forthcoming chapters show that the modeling of phenomena in life sciences may require the above particularizations. For instance, the models of social dynamics analyzed in Chapter 5 are such that localization of interacting individuals is not relevant in assessing their state described by the level of wealth. On the other hand, modeling the interplay between migration phenomena and social dynamics requires including, in the microscopic state, a space variable related to regional areas. Migration from one area to the other is related not only to interactions, but also to the social state.

2.3 Modeling Microscopic Interactions

The first step towards the derivation of evolution equations for the distribution function defined in Section 2.2 is the modeling of microscopic interactions. This section develops a mathematical framework suitable for including a large variety of specific interaction models. The contents of this section essentially refer to the papers by Bellouquid and Delitala (2004), (2005), where the derivation of the above-mentioned mathematical framework was developed for multicellular biological systems. The same reasoning can also be applied in the relatively more general cases dealt with in this book. The derivation refers to two type of interactions.

• *Short range binary interactions* which refer to the mutual actions between the *test* (or *candidate*) and the *field* active particle, when the field particle enters into the short range interaction domain of the test (or candidate) particle. Such a domain is of the same order of the size of the interacting active particles.

• *Long range mean field interactions* which refer to the action over the *test* active particles applied by all *field* active particles which are in the long range action domain Ω of the field particle. The action is still of the type of binary encounters.

Short and long range interactions are visualized in Figures 2.3.1 and 2.3.2, respectively. Moreover, for both types of interactions, we consider the following classifications.

• *Conservative interactions* which modify the state, mechanical and/or activity, of the interacting active particles, but not the size of the population.

• *Proliferative or destructive interactions* which generate death or birth of active particles due to pair interactions.

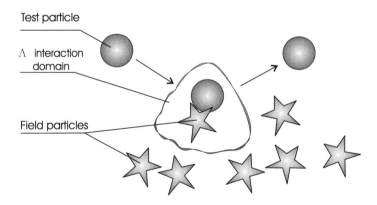

Fig. 2.3.1: Short range interactions.

The dynamics of conservative interactions is visualized in Figure 2.3.3, while proliferative and destructive interactions are visualized in Figures 2.3.4 and 2.3.5, respectively.

The analysis of the two types of modeling is developed in the two subsections which follow, while the last subsection provides a critical analysis.

2.3.1 Short Range Interactions

Let us first consider *localized conservative interactions* between the *test* or *candidate* active particle with state $\mathbf{w}_*$ and the *field* active particle

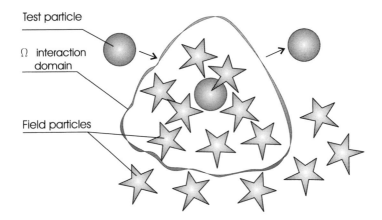

Fig. 2.3.2: Long range interactions.

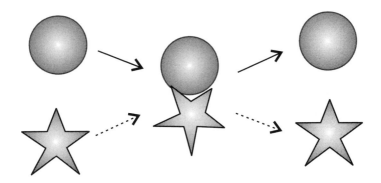

Fig. 2.3.3: Conservative interactions.

with state $\mathbf{w}^*$. The modeling of microscopic interactions can be based on the knowledge of the following two quantities:

- The **encounter rate**, which depends for each pair on the relative velocity

$$\eta = c_0 \, \delta(\mathbf{x}_* - \mathbf{x}^*) \, |\mathbf{v}_* - \mathbf{v}^*| \,, \qquad (2.3.1)$$

where c_0 is a constant and δ is Dirac's function;

and

- the **transition probability density** function

$$\varphi(\mathbf{w}_*, \mathbf{w}^*; \mathbf{w}) : \quad D_{\mathbf{w}} \times D_{\mathbf{w}} \times D_{\mathbf{w}} \ \to \ \mathbb{R}_+ \,, \qquad (2.3.2)$$

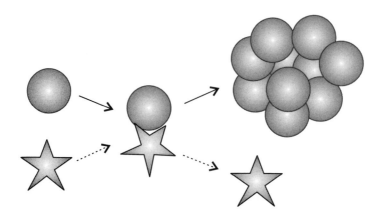

Fig. 2.3.4: Proliferative interactions.

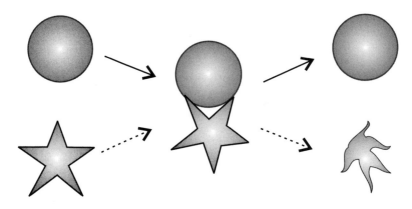

Fig. 2.3.5: Destructive interactions.

which denotes the probability density that a **_candidate_** active particle
with state $\mathbf{w}_*$ falls into the state $\mathbf{w}$ of the test particle after an interaction with a field particle with state $\mathbf{w}^*$. The interaction term φ has the
structure of a probability density function with respect to the variable
$\mathbf{w}$, which describes the microscopic state

$$\forall \, \mathbf{w}_*, \mathbf{w}^* : \quad \int_{D_{\mathbf{w}}} \varphi(\mathbf{w}_*, \mathbf{w}^*; \mathbf{w}) \, d\mathbf{w} = 1 \,. \qquad (2.3.3)$$

Remark 2.3.1. *Specific models can be practically designed assuming that*
φ *is given by the product of the transition densities* $\mathcal{M}$ *related to the*
mechanical variable with transition density $\mathcal{B}$ *of the activity:*

$$\varphi(\mathbf{w}_*, \mathbf{w}^*; \mathbf{w}) = \mathcal{M}(\mathbf{v}_*, \mathbf{v}^*; \mathbf{v}|\mathbf{u}_*, \mathbf{u}^*)\delta(\mathbf{x} - \mathbf{x}_*)\mathcal{B}(\mathbf{u}_*, \mathbf{u}^*; \mathbf{u}) \,, \qquad (2.3.4)$$

where the velocity dynamics is conditioned by the activities of the interacting pairs.

Let us now consider *localized nonconservative interactions* between test and field active particles which occur with the above-defined encounter rate. Proliferative and/or destructive encounters can be described by the following quantity:

- The *source/sink short range distribution function*

$$\psi(\mathbf{w}_*, \mathbf{w}^*; \mathbf{w}) = \mu(\mathbf{u}_*, \mathbf{u}^*)\delta(\mathbf{w} - \mathbf{w}_*), \qquad (2.3.5)$$

where μ is the *proliferation (or destruction) rate* generated by the interaction. This framework is based on the assumption that this term depends only on the activities of the interacting pair: proliferative and destructive processes occur in the microscopic state of the test particle.

The generalizations to systems of populations is immediate. The quantities to be defined are as follows:

- The *encounter rate*: $\eta_{ij}(\mathbf{w}_*, \mathbf{w}^*)$ depending both on the states and on the type of populations of the interacting pairs. The model

$$\eta_{ij}(\mathbf{w}_*, \mathbf{w}^*) = c_{ij}\,\delta(\mathbf{x}_* - \mathbf{x}^*)\,|\mathbf{v}_* - \mathbf{v}^*|, \qquad (2.3.6)$$

where c_{ij} is a constant for each pair of populations, corresponds to (2.3.1.);

- The *transition probability density* function $\varphi_{ij}(\mathbf{w}_*, \mathbf{w}^*; \mathbf{w})$, which has the structure of a probability density function with respect to the variable $\mathbf{w}$, is such that $\varphi_{ij}(\mathbf{w}_*, \mathbf{w}^*; \mathbf{w})$ denotes the probability density that a test particle with state $\mathbf{w}_*$ belonging to the ith population falls into the state $\mathbf{w}$ after an interaction with a field particle belonging to the jth population with state $\mathbf{w}^*$. In particular, as in model (2.3.4), φ_{ij} can be assumed to be given by the product of the transition densities related to the mechanical variable with the one of the activity,

$$\varphi_{ij} = \mathcal{M}_{ij}(\mathbf{v}_*, \mathbf{v}^*; \mathbf{v}|\mathbf{u}_*, \mathbf{u}^*)\delta(\mathbf{x} - \mathbf{x}_*)\mathcal{B}_{ij}(\mathbf{u}_*, \mathbf{u}^*; \mathbf{u}), \qquad (2.3.7)$$

where the output of the mechanical interactions depends on the input velocity and activities only, while interactions involving the activities depend on their input states only;

- The *source/sink short range distribution function*

$$\psi_{ij}(\mathbf{w}_*, \mathbf{w}^*; \mathbf{w}) = \mu_{ij}(\mathbf{w}_*, \mathbf{w}^*)\delta(\mathbf{w} - \mathbf{w}_*), \qquad (2.3.8)$$

where μ_{ij} is the proliferation (or destruction) rate generated by the interaction of the test active particle, with state $\mathbf{w}_*$, belonging to the ith population with a field active particle, with state $\mathbf{w}^*$, belonging to the jth population. Proliferative and destructive events occur in the microscopic state of the test active particle. Again, as in model (2.3.5), the term μ_{ij} can be assumed to depend on the activities only,

$$\mu_{ij} = \mu_{ij}(\mathbf{u}_*, \mathbf{u}^*). \qquad (2.3.9)$$

Remark 2.3.2. *The preceding particularizations are essentially based on the assumption that interactions that modify the activity variable are affected by mechanical interactions only through the encounter rate, while mechanical interactions are affected by the activity state. Namely, the active particles select a strategy to move within their environment based on the activity state of the interacting pair. The output of the interaction is assumed to be localized in the same point of the test active particle according to the assumption of short range interactions.*

If the quantities above are known, it is possible to compute the fluxes, related to both types of interactions (conservative and proliferative or destructive) in the elementary volume $[\mathbf{w}, \mathbf{w} + d\mathbf{w}]$ of the space of microscopic states. The calculations reported, in the following paragraphs are used for the derivation of the evolution equations in Section 2.4. Calculations refer directly to the case of systems of several populations; the relatively simple case of one population only is easily obtained by eliminating interactions with other populations.

Still referring to localized interactions, the inlet flux in the elementary volume of the state space due to conservative interactions is given as follows:

$$\mathcal{C}_i^+[\mathbf{f}](t, \mathbf{x}, \mathbf{v}, \mathbf{u}) = \sum_{j=1}^{n} \int_{(D_\mathbf{v} \times D_\mathbf{u})^2} c_{ij} |\mathbf{v}_* - \mathbf{v}^*| \mathcal{M}_{ij}(\mathbf{v}_*, \mathbf{v}^*; \mathbf{v} | \mathbf{u}_*, \mathbf{u}^*)$$
$$\times \mathcal{B}_{ij}(\mathbf{u}_*, \mathbf{u}^*; \mathbf{u}) f_i(t, \mathbf{x}, \mathbf{v}_*, \mathbf{u}_*)$$
$$\times f_j(t, \mathbf{x}, \mathbf{v}^*, \mathbf{u}^*) \, d\mathbf{v}_* \, d\mathbf{v}^* \, d\mathbf{u}_* \, d\mathbf{u}^*, \qquad (2.3.10)$$

while the outlet flux is

$$\mathcal{C}_i^-[\mathbf{f}](t, \mathbf{x}, \mathbf{v}, \mathbf{u}) = f_i(t, \mathbf{x}, \mathbf{v}, \mathbf{u}) \sum_{j=1}^{n} \int_{D_\mathbf{v} \times D_\mathbf{u}} c_{ij} |\mathbf{v} - \mathbf{v}^*|$$
$$\times f_j(t, \mathbf{x}, \mathbf{v}^*, \mathbf{u}^*) \, d\mathbf{v}^* \, d\mathbf{u}^*. \qquad (2.3.11)$$

Analogous calculations, related to proliferative or destructive interactions, yield

$$
\mathcal{I}_i[\mathbf{f}](t, \mathbf{x}, \mathbf{v}, \mathbf{u}) = f_i(t, \mathbf{x}, \mathbf{v}, \mathbf{u}) \sum_{j=1}^{n} \int_{D_{\mathbf{v}} \times D_{\mathbf{u}}} c_{ij} |\mathbf{v} - \mathbf{v}^*| \mu_{ij}(\mathbf{u}, \mathbf{u}^*)
$$

$$
\times f_j(t, \mathbf{x}, \mathbf{v}^*, \mathbf{u}^*) \, d\mathbf{v}^* \, d\mathbf{u}^* \,. \tag{2.3.12}
$$

2.3.2 Long Range Interactions

Let us now consider the modeling of *long range conservative interactions* between the *test* active particle with state $\mathbf{w}_*$ and *field* active particle with state $\mathbf{w}^*$. The microscopic modeling of pair interactions, conservative and proliferative/destructive, can be respectively based on the knowledge of the following quantities:

- The *long range vector action* $\mathcal{P} = \mathcal{P}(\mathbf{x}, \mathbf{x}_*, \mathbf{v}, \mathbf{v}_*, \mathbf{u}, \mathbf{u}_*)$ over the *test* active particle with microscopic state $\mathbf{w}$ due to the *field* active particle with state $\mathbf{w}_*$, where $\mathcal{P}$ is a vector with the same dimension of the microscopic state $\{\mathbf{v}, \mathbf{u}\}$.

- The *source/sink long range term* $\sigma(\mathbf{w}, \mathbf{w}_*)$, which denotes the proliferative or destructive rate related to an active particle with state $\mathbf{w}$ with a birth or death process in its state due to the interaction with the field entity with state $\mathbf{w}_*$.

Remark 2.3.3. *Models of conservative interactions can be practically designed assuming that $\mathcal{P}$ can be split into the sum of the vectors $\mathcal{P}^m$ acting over $\mathbf{v}$, and $\mathcal{P}^a$ acting over $\mathbf{u}$. Moreover, the additional assumptions $\mathcal{P}^m = \mathcal{P}^m(\mathbf{x}, \mathbf{x}_*; \mathbf{u}, \mathbf{u}_*)$, and $\mathcal{P}^a = \mathcal{P}^a(\mathbf{u}, \mathbf{u}_*)$ can be proposed for practical applications.*

Remark 2.3.4. *Specific models of nonconservative interactions can be practically designed assuming that proliferative and destructive rates depend only on the activity: $\sigma = \sigma(\mathbf{u}, \mathbf{u}_*)$.*

Again, the generalization to the case of several interacting populations is straightforward. Simply, each quantity has to be labeled by the subscripts "ij", for $i, j = 1, \ldots, n$, to denote interactions between active particles of the ith and jth populations. In detail, the various terms which describe microscopic localized interactions are the following:

- The *long range vector action* $\mathcal{P}_{ij} = \mathcal{P}_{ij}(\mathbf{x}, \mathbf{x}_*, \mathbf{v}, \mathbf{v}_*, \mathbf{u}, \mathbf{u}_*)$ over the *test* active particle (of the ith population) with microscopic state $\mathbf{w}$ due to the *field* active particle (of the jth population) with $\mathbf{w}_*$.

- The *proliferative/destructive long range term* $\sigma_{ij}(\mathbf{w}, \mathbf{w}_*)$, which denotes the proliferative or destructive rate related to the test active particle with state $\mathbf{w}$ belonging to the ith population with a birth or death process in its state due to the interaction with the field active particle with state $\mathbf{w}_*$ belonging to the jth population.

Again, it is useful to specialize the above term as described in Remarks 2.3.3 and 2.3.4. Specifically, the vector action $\mathcal{P}_{ij}$ can be split into the vectors $\mathcal{P}_{ij}^m$ acting over the mechanical variables $\mathbf{v}$, and $\mathcal{P}_{ij}^a$ acting over the activity variable $\mathbf{u}$, with

$$\mathcal{P}_{ij}^m = \mathcal{P}_{ij}^m(\mathbf{x}, \mathbf{x}_*; \mathbf{u}, \mathbf{u}_*), \qquad \text{and} \qquad \mathcal{P}_{ij}^a = \mathcal{P}_{ij}^a(\mathbf{u}, \mathbf{u}_*),$$

while the proliferative/destructive term depends on the activity variables only: $\sigma_{ij} = \sigma_{ij}(\mathbf{u}, \mathbf{u}_*)$.

The knowledge of the above quantity allows the calculation of the **resultant action** of all active particles of the jth populations in the action domain Ω of the test particle 0f the ith population:

$$\mathcal{F}_{ij}[\mathbf{f}](t, \mathbf{x}, \mathbf{v}, \mathbf{u}) = \int_{\mathcal{D}} \mathcal{P}_{ij}(\mathbf{x}, \mathbf{x}_*, \mathbf{v}, \mathbf{v}_*, \mathbf{u}, \mathbf{u}_*)$$
$$\times f_j(t, \mathbf{x}_*, \mathbf{v}_*, \mathbf{u}_*)\, d\mathbf{x}_*\, d\mathbf{v}_*\, d\mathbf{u}_*, \qquad (2.3.13)$$

where $\mathcal{D} = \Omega \times D_\mathbf{v} \times D_\mathbf{u}$, and where Ω is the interaction domain of the test active particle: if $\mathbf{x}_* \notin \Omega \Rightarrow \mathcal{P}_{ij} = 0$.

Referring to nonconservative interactions, pair interactions generate a proliferation and/or destruction in the state $\mathbf{w}$ described by the following term:

$$\sigma_{ij} = \sigma_{ij}(\mathbf{x}, \mathbf{x}_*, \mathbf{v}, \mathbf{v}_*, \mathbf{u}, \mathbf{u}_*).$$

This microscopic quantity allows the calculation of the following source/sink term:

$$\mathcal{S}_{ij}[\mathbf{f}](t, \mathbf{x}, \mathbf{v}, \mathbf{u}) = \int_{\mathcal{D}} \sigma_{ij}(\mathbf{x}, \mathbf{x}_*, \mathbf{v}, \mathbf{v}_*, \mathbf{u}, \mathbf{u}_*)$$
$$\times f_j(t, \mathbf{x}_*, \mathbf{v}_*, \mathbf{u}_*)\, d\mathbf{x}_*\, d\mathbf{v}_*\, d\mathbf{u}_*, \qquad (2.3.14)$$

which is somehow analogous to the one related to short term interactions. The difference is related to the finite long-range interaction domain Ω, while σ_{ij} also depends on the positions $\mathbf{x}$ and $\mathbf{x}_*$ to describe the decay in space of the proliferative/destructive actions.

These general results can be specialized as follows:

$$\mathcal{F}_{ij}^m[\mathbf{f}](t, \mathbf{x}) = \int_{\mathcal{D}} \mathcal{P}_{ij}^m(\mathbf{x}, \mathbf{x}_*, \mathbf{u}, \mathbf{u}_*)\, f_j(t, \mathbf{x}_*, \mathbf{v}_*, \mathbf{u}_*)\, d\mathbf{x}_*\, d\mathbf{v}_*\, d\mathbf{u}_*, \qquad (2.3.15)$$

$$\mathcal{F}_{ij}^a[\mathbf{f}](t,\mathbf{u}) = \int_{\mathcal{D}} \mathcal{P}_{ij}^a(\mathbf{u},\mathbf{u}_*)\, f_j(t,\mathbf{x}_*,\mathbf{v}_*,\mathbf{u}_*)\, d\mathbf{x}_*\, d\mathbf{v}_*\, d\mathbf{u}_* , \qquad (2.3.16)$$

and

$$\mathcal{S}_{ij}[\mathbf{f}](t,\mathbf{u}) = \int_{\mathcal{D}} \sigma_{ij}(\mathbf{u},\mathbf{u}_*)\, f_j(t,\mathbf{x}_*,\mathbf{v}_*,\mathbf{u}_*)\, d\mathbf{x}_*\, d\mathbf{v}_*\, d\mathbf{u}_* . \qquad (2.3.17)$$

The transport terms for each population can be computed, according to Eqs. (2.3.15) and (2.3.16), as follows:

$$T_i^m[\mathbf{f}](t,\mathbf{x},\mathbf{u}) = \sum_{j=1}^{n} \mathcal{F}_{ij}^m[\mathbf{f}](t,\mathbf{x},\mathbf{u}) \cdot \nabla_{\mathbf{v}} f_i(t,\mathbf{x},\mathbf{v},\mathbf{u}) , \qquad (2.3.18)$$

and

$$T_i^a[\mathbf{f}](t,\mathbf{x},\mathbf{u}) = \sum_{j=1}^{n} \nabla_{\mathbf{u}} \cdot \left(\mathcal{F}_{ij}^a[\mathbf{f}](t,\mathbf{u}) f_i(t,\mathbf{x},\mathbf{v},\mathbf{u}) \right) , \qquad (2.3.19)$$

while the proliferative destructive term is given by

$$S_i[\mathbf{f}](t,\mathbf{u}) = \sum_{j=1}^{n} \mathcal{S}_{ij}[\mathbf{f}](t,\mathbf{u}) . \qquad (2.3.20)$$

2.3.3 Additional Analysis

The preceding two subsections have derived suitable mathematical frameworks for the modeling of microscopic interactions of active particles. They must be considered as mathematical structures as the whole modeling process is completed only if detailed expressions of the interaction terms are properly designed with reference to the specific physical system under consideration. This means modeling detailed expressions of the terms η_{ij}, φ_{ij}, and ψ_{ij} in the case of short range interactions, and $\mathcal{P}_{ij}^m$, $\mathcal{P}_{ij}^a$, and σ_{ij} in the case of long range applications. Only if these terms are modeled on the basis of a theoretical or phenomenological approach, is the whole modeling process finally complete.

The mathematical frameworks above can be generalized or even improved. For instance, the modeling can include both short and long range interactions, or one can consider the case of proliferation in a population different from the one of the interacting pairs. These generalizations will be discussed in Chapters 3 and 4.

Once more, we stress one of the differences between active and classical particles: an active particle feels the presence of another particle at a distance, where as classical particles "feel" each other only when mutual long distance forces, e.g., electromagnetic, are exchanged. Moreover, the type of active particle interactions may even differ with the distance, e.g., proliferative/destructive encounters occur at short distances, while conservative encounters are effective at long distances.

2.4 Mathematical Frameworks

The mathematical models of interactions at the microscopic scale described in the preceding section are a preliminary step to the derivation of evolution equations, which are used as a formal framework to be properly specialized with reference to specific physical systems. In other words, a detailed analysis (and modeling) of individual interactions generates models suitable to describe the evolution of a class of large systems of interacting active particles.

The derivation needs to be particularized for each class of microscopic interactions. Let us first consider the case of **models with short range interactions**. The evolution equation is obtained, similarly to the case of the Boltzmann equation, by equating the local rate of increase of particles, with state $\mathbf{w}$ in the elementary volume $[\mathbf{w}, \mathbf{w} + d\mathbf{w}]$ around $\mathbf{w}$, due to the net flux of active particles which reach such a state due to interactions and to proliferation in the same state minus the outlet due to conservative interactions and destruction into such a state. The scheme, in the absence of external action, is as follows:

$$\frac{\partial f}{\partial t} + \mathbf{v} \cdot \nabla_{\mathbf{x}} f = J[f] = \mathcal{C}^+[f] - \mathcal{C}^-[f] + \mathcal{I}[f] \,, \qquad (2.4.1)$$

which corresponds to the balance equation described in the flow chart of Figure 2.4.1 for $f = f(t, \mathbf{x}, \mathbf{v}, \mathbf{u})$.

Consequently, the formal expressions given in Eqs. (2.3.10)–(2.3.12) can be finally used to derive the following class of evolution equations:

$$\left(\frac{\partial}{\partial t} + \mathbf{v} \cdot \nabla_{\mathbf{x}}\right) f(t, \mathbf{x}, \mathbf{v}, \mathbf{u}) = \int_{D \times D} c \, |\mathbf{v}_* - \mathbf{v}^*| \mathcal{M}(\mathbf{v}_*, \mathbf{v}^*, \mathbf{v} | \mathbf{u}_*, \mathbf{u}^*)$$

$$\times \, \mathcal{B}(\mathbf{u}_*, \mathbf{u}^*; \mathbf{u}) f(t, \mathbf{x}, \mathbf{v}_*, \mathbf{u}_*) f(t, \mathbf{x}, \mathbf{v}^*, \mathbf{u}^*) \, d\mathbf{v}_* \, d\mathbf{u}_* \, d\mathbf{v}^* \, d\mathbf{u}^*$$

$$- \, f(t, \mathbf{x}, \mathbf{v}, \mathbf{u}) \int_D c \, |\mathbf{v} - \mathbf{v}^*| f(t, \mathbf{x}, \mathbf{v}^*, \mathbf{u}^*) \, d\mathbf{v}^* \, d\mathbf{u}^*$$

$$+ \, f(t, \mathbf{x}, \mathbf{v}, \mathbf{u}) \int_D c \, |\mathbf{v} - \mathbf{v}^*| \mu(\mathbf{u}, \mathbf{u}^*) f(t, \mathbf{x}, \mathbf{v}^*, \mathbf{u}^*) \, d\mathbf{v}^* \, d\mathbf{u}^* \,, \qquad (2.4.2)$$

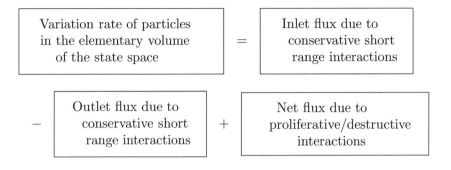

Fig. 2.4.1: Mass balance in the state space for short range interactions.

where $D = D_{\mathbf{v}} \times D_{\mathbf{u}}$.

The generalization to several interacting populations is immediate. It refers to the following balance equation:

$$\frac{df_i}{dt} = J_i[\mathbf{f}] = \sum_{j=1}^{n} J_{ij}[\mathbf{f}] = \mathcal{C}_i^+[\mathbf{f}] - \mathcal{C}_i^-[\mathbf{f}] + \mathcal{I}_i[\mathbf{f}], \qquad (2.4.3)$$

where the interaction terms correspond to the sum of interactions with all populations. Therefore, using (2.3.1)–(2.3.12) the following structure is obtained:

$$\left(\frac{\partial}{\partial t} + \mathbf{v} \cdot \nabla_{\mathbf{x}}\right) f_i(t, \mathbf{x}, \mathbf{v}, \mathbf{u}) = J_i[\mathbf{f}](t, \mathbf{x}, \mathbf{v}, \mathbf{u}) = \sum_{j=1}^{n} J_{ij}[\mathbf{f}](t, \mathbf{x}, \mathbf{v}, \mathbf{u})$$

$$= \sum_{j=1}^{n} \int_{D \times D} c_{ij} |\mathbf{v}_* - \mathbf{v}^*| \mathcal{M}_{ij}(\mathbf{v}_*, \mathbf{v}^*, \mathbf{v} | \mathbf{u}_*, \mathbf{u}^*)$$

$$\times \mathcal{B}_{ij}(\mathbf{u}_*, \mathbf{u}^*; \mathbf{u}) f_i(t, \mathbf{x}, \mathbf{v}_*, \mathbf{u}_*) f_j(t, \mathbf{x}, \mathbf{v}^*, \mathbf{u}^*) \, d\mathbf{v}_* \, d\mathbf{u}_* \, d\mathbf{v}^* \, d\mathbf{u}^*$$

$$- f_i(t, \mathbf{x}, \mathbf{v}, \mathbf{u}) \sum_{j=1}^{n} \int_{D} c_{ij} |\mathbf{v} - \mathbf{v}^*| f_j(t, \mathbf{x}, \mathbf{v}^*, \mathbf{u}^*) \, d\mathbf{v}^* \, d\mathbf{u}^*$$

$$+ f_i(t, \mathbf{x}, \mathbf{v}, \mathbf{u}) \sum_{j=1}^{n} \int_{D} c_{ij} |\mathbf{v} - \mathbf{v}^*| \mu_{ij}(\mathbf{u}, \mathbf{u}^*)$$

$$\times f_j(t, \mathbf{x}, \mathbf{v}^*, \mathbf{u}^*) \, d\mathbf{v}^* \, d\mathbf{u}^* . \qquad (2.4.4)$$

Let us now refer to **models with long range interactions**. The evolution equation is obtained following the scheme described in the flow chart

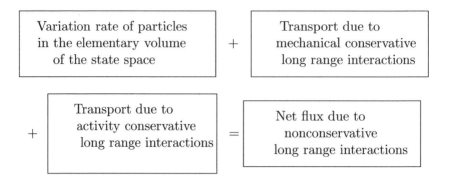

Fig. 2.4.2: Mass balance for long range interactions.

of Figure 2.4.2. In the absence of external action, the evolution equation is

$$\frac{\partial f}{\partial t} + \mathbf{v} \cdot \nabla_{\mathbf{x}} f + T^m[f] + T^a[f] = S[f].$$ (2.4.5)

Using (2.3.13)–(2.3.20) yields

$$\frac{\partial}{\partial t} f(t, \mathbf{x}, \mathbf{v}, \mathbf{u}) + \mathbf{v} \cdot \nabla_{\mathbf{x}} f(t, \mathbf{x}, \mathbf{v}, \mathbf{u})$$

$$+ \left(\int_{\mathcal{D}} \mathcal{P}^m(\mathbf{x}, \mathbf{x}_*, \mathbf{u}, \mathbf{u}_*) f(t, \mathbf{x}_*, \mathbf{v}_*, \mathbf{u}_*) \, d\mathbf{x}_* \, d\mathbf{v}_* \, d\mathbf{u}_* \right) \cdot \nabla_{\mathbf{v}} f(t, \mathbf{x}, \mathbf{v}, \mathbf{u})$$

$$+ \nabla_{\mathbf{u}} \cdot \left(f(t, \mathbf{x}, \mathbf{v}, \mathbf{u}) \int_{\mathcal{D}} \mathcal{P}^a(\mathbf{x}, \mathbf{x}_*, \mathbf{u}, \mathbf{u}_*) f(t, \mathbf{x}_*, \mathbf{v}_*, \mathbf{u}_*) \, d\mathbf{x}_* \, d\mathbf{v}_* \, d\mathbf{u}_* \right)$$

$$= f(t, \mathbf{x}, \mathbf{v}, \mathbf{u}) \int_{\mathcal{D}} \sigma(\mathbf{u}, \mathbf{u}_*) f(t, \mathbf{x}_*, \mathbf{v}_*, \mathbf{u}_*) \, d\mathbf{x}_* \, d\mathbf{v}_* \, d\mathbf{u}_*.$$ (2.4.6)

The generalization to several interacting populations is obtained by considering interactions with all populations. It corresponds to the following scheme:

$$\frac{\partial f_i}{\partial t} + \mathbf{v} \cdot \nabla_{\mathbf{x}} f_i + T_i^m[\mathbf{f}] + T_i^a[\mathbf{f}] = \mathcal{S}_i[\mathbf{f}],$$ (2.4.7)

where explicitly,

$$\frac{\partial}{\partial t} f_i(t, \mathbf{x}, \mathbf{v}, \mathbf{u}) + \mathbf{v} \cdot \nabla_{\mathbf{x}} f_i(t, \mathbf{x}, \mathbf{v}, \mathbf{u})$$

$$+ \sum_{j=1}^{n} \mathcal{F}_{ij}^m[\mathbf{f}](t, \mathbf{x}, \mathbf{u}) \cdot \nabla_{\mathbf{v}} f_i(t, \mathbf{x}, \mathbf{v}, \mathbf{u})$$

$$+ \nabla_{\mathbf{u}} \cdot \left(f_i(t, \mathbf{x}, \mathbf{v}, \mathbf{u}) \sum_{j=1}^{n} \mathcal{F}_{ij}^a[\mathbf{f}](t, \mathbf{u}) \right)$$

$$= f_i(t, \mathbf{x}, \mathbf{v}, \mathbf{u}) \sum_{j=1}^{n} \int_{\mathcal{D}} \sigma_{ij}(\mathbf{u}, \mathbf{u}_*)$$

$$\times f_j(t, \mathbf{x}_*, \mathbf{v}_*, \mathbf{u}_*) \, d\mathbf{x}_* \, d\mathbf{v}_* \, d\mathbf{u}_* \,. \tag{2.4.8}$$

Active particles may be subject to an external field $\mathbf{F}(t, \mathbf{x})$ acting over the velocity variable and/or to an action $\mathbf{K}(t, \mathbf{x})$ acting over the activity variable. Therefore, the left-hand side term should include the corresponding transport terms. The structure of the equations, for one population only, is as follows:

$$\frac{\partial f}{\partial t} + \mathbf{v} \cdot \nabla_{\mathbf{x}} f + \mathbf{F} \cdot \nabla_{\mathbf{v}} f + \mathbf{K} \cdot \nabla_{\mathbf{u}} f = J[f], \tag{2.4.9}$$

and

$$\frac{\partial f}{\partial t} + \mathbf{v} \cdot \nabla_{\mathbf{x}} f + \mathbf{F} \cdot \nabla_{\mathbf{v}} f + \mathbf{K} \cdot \nabla_{\mathbf{u}} f + T^m[\mathbf{f}] + T^a[\mathbf{f}] = \mathcal{S}[\mathbf{f}], \tag{2.4.10}$$

while the generalization to several interacting populations is immediate.

2.5 Some Particular Frameworks

This section reports some technical particularizations, among various conceivable ones, referred to the mathematical framework derived in Section 2.4. The following special cases are considered:

i) Models of systems characterized by dominant activity interactions, while the distribution over the mechanical microscopic state is uniform in space and constant in time;

ii) Models of systems characterized by dominant mechanical interactions, while the distribution over the activity state is everywhere constant in time;

iii) Models of systems where the microscopic state is limited to the activity variable only.

The derivation of specialized frameworks is developed for both types of interactions: short and long range. The derivation is developed in the absence of external actions. Inserting these additional terms is simply a matter of a few more calculations.

2.5.1 Models with Dominant Activity Interactions

The class of models proposed in Section 2.4 can be simplified, in some cases, with reference to physical situations where some specific phenomena are less relevant (or are negligible) with respect to other ones. This is the case of systems in spatial homogeneity with a constant in time distribution over the velocity variable. The evolution equation is obtained by integrating over the domain of the velocity variable.

Let us now assume:

$$f_i(t, \mathbf{v}, \mathbf{u}) = f_i^a(t, \mathbf{u}) P_i(\mathbf{v}),$$ (2.5.1)

uniform in the space variable, where $P_i(\mathbf{v})$ is such that

$$\int_{D_\mathbf{v}} P_i(\mathbf{v}) d\mathbf{v} = 1.$$

Using the above assumption, deal with mechanical interactions, under the technical assumption that $\mathcal{M}_{ij}$ is replaced by

$$\mathcal{M}_{ij}^m(\mathbf{v}_*, \mathbf{v}^*; \mathbf{v}(\mathbf{u}_*, \mathbf{u}^*)) = \delta(\mathbf{v}_* - \mathbf{v}).$$ (2.5.2)

Integration over the velocity variable yields

$$\frac{\partial}{\partial t} f_i^a(t, \mathbf{u}) = J_i[\mathbf{f}^a](t, \mathbf{u})$$

$$= \sum_{j=1}^n \eta_{ij}^0 \int_{D_\mathbf{u} \times D_\mathbf{u}} \mathcal{B}_{ij}(\mathbf{u}_*, \mathbf{u}^*; \mathbf{u}) f_i^a(t, \mathbf{u}_*) f_j^a(t, \mathbf{u}^*) \, d\mathbf{u}_* \, d\mathbf{u}^*$$

$$- f_i^a(t, \mathbf{u}) \sum_{j=1}^n \eta_{ij}^0 \int_{D_\mathbf{u}} [1 - \mu_{ij}(\mathbf{u}, \mathbf{u}^*)] f_j^a(t, \mathbf{u}^*) \, d\mathbf{u}^*,$$ (2.5.3)

where

$$\eta_{ij}^0 = \int_{D_\mathbf{v} \times D_\mathbf{v}} c_{ij} |\mathbf{v} - \mathbf{v}^*| P_j(\mathbf{v}^*) P_i(\mathbf{v}) \, d\mathbf{v}^* \, d\mathbf{v}.$$ (2.5.4)

The same analysis can be developed for long range interactions. In this case, the evolution equation is written as

$$\frac{\partial f_i^a}{\partial t}(t, \mathbf{u}) + \nabla_\mathbf{u} \cdot \left(\sum_{j=1}^n \mathcal{F}_{ij}^a[f^a] f_i^a \right)(t, \mathbf{u})$$

$$= f_i^a(t, \mathbf{u}) \sum_{j=1}^n \int_{D_\mathbf{u}} \sigma_{ij}(\mathbf{u}, \mathbf{u}_*) f_j^a(t, \mathbf{u}_*) \, d\mathbf{u}_*,$$ (2.5.5)

where

$$\mathcal{F}_{ij}^a = \int_{D_\mathbf{u}} \mathcal{P}_{ij}^a(\mathbf{u}, \mathbf{u}_*) f_j^a(t, \mathbf{u}_*) \, d\mathbf{u}_* \,. \tag{2.5.6}$$

The mathematical structures reported in (2.5.3)–(2.5.6) are based on the assumption that $P_i(\mathbf{v})$ depends on the velocity variable only. More generally, it can depend on macroscopic variables as in the case of the Boltzmann equation. Therefore, the encounter rate is not a constant, but can be referred to macroscopic quantities, as we shall see when we study the specific applications in the second part of the book.

2.5.2 Models with Dominant Mechanical Interactions

An analogous reasoning can be applied to the case of dominant mechanical interactions. This involves systems with an everywhere constant in time distribution over the activity. In other words, the rules of the mechanical interactions depend on the activity states of the interacting pair. However, the distribution over the activity state is not influenced by interactions. Consequently, the evolution equation refers to the mechanical variable obtained by integrating over the domain of the activity variable. The calculations are analogous to those we have seen above. The assumption needed to derive the equation is as follows:

$$f_i(t, \mathbf{x}, \mathbf{v}, \mathbf{u}) = f_i^m(t, \mathbf{x}, \mathbf{v}) P_i(\mathbf{u}), \qquad \int_{D_\mathbf{u}} P_i(\mathbf{u}) d\mathbf{u} = 1, \tag{2.5.7}$$

where the proliferation term is assumed to be equal to zero: $\mu_{ij} = 0$.

Therefore, substituting into (2.4.4) and averaging over the activity variable yields

$$\frac{\partial}{\partial t} f_i^m(t, \mathbf{x}, \mathbf{v}) + \mathbf{v} \cdot \nabla_\mathbf{x} f_i^m(t, \mathbf{x}, \mathbf{v}) = J_i[\mathbf{f}^m](t, \mathbf{x}, \mathbf{v})$$

$$= \sum_{j=1}^n \int_{D_\mathbf{v} \times D_\mathbf{v}} c_{ij} |\mathbf{v}_* - \mathbf{v}^*| \mathcal{M}_{ij}^m(\mathbf{v}_*, \mathbf{v}^*; \mathbf{v})$$

$$\times f_i^m(t, \mathbf{x}, \mathbf{v}_*) f_j^m(t, \mathbf{x}, \mathbf{v}^*) d\mathbf{v}_* \, d\mathbf{v}^*$$

$$- f_i^m(t, \mathbf{x}, \mathbf{v}) \sum_{j=1}^n \int_{D_\mathbf{v}} c_{ij} |\mathbf{v} - \mathbf{v}^*| f_j^m(t, \mathbf{x}^*, \mathbf{v}^*) d\mathbf{v}^* \,, \tag{2.5.8}$$

where

$$\mathcal{M}_{ij}^m(\mathbf{v}_*, \mathbf{v}^*; \mathbf{v}) = \int_{D_\mathbf{u} \times D_\mathbf{u}} \mathcal{M}_{ij}(\mathbf{v}_*, \mathbf{v}^*; \mathbf{v} | \mathbf{u}_*, \mathbf{u}^*)$$

$$\times P_i(\mathbf{u}_*) P_j(\mathbf{u}^*) \, d\mathbf{u}_* \, d\mathbf{u}^* \,. \tag{2.5.9}$$

The same analysis can be developed for long range interactions. The evolution equation is obtained assuming $\sigma_{ij} = \mathcal{P}_{ij}^{b} = 0$ and averaging over the activity variable. Moreover, assuming that the mechanical interaction term is positional, we obtain

$$\frac{\partial}{\partial t} f_i^m(t, \mathbf{x}, \mathbf{v}) + \mathbf{v} \cdot \nabla_{\mathbf{x}} f_i^m(t, \mathbf{x}, \mathbf{v})$$

$$+ \sum_{j=1}^{n} \left[\int_{D_{\mathbf{x}} \times D_{\mathbf{v}}} \mathcal{P}_{ij}^{mm}(\mathbf{x}, \mathbf{x}_*) f_j^m(t, \mathbf{x}_*, \mathbf{v}_*) \, d\mathbf{x}_* \, d\mathbf{v}_* \right]$$

$$\cdot \nabla_{\mathbf{v}} f_i^m(t, \mathbf{x}, \mathbf{v}) = 0, \tag{2.5.10}$$

where

$$\mathcal{P}_{ij}^{mm}(\mathbf{x}, \mathbf{x}_*) = \int_{D_{\mathbf{u}} \times D_{\mathbf{u}}} \mathcal{P}_{ij}^{m}(\mathbf{x}, \mathbf{x}_*, \mathbf{u}, \mathbf{u}_*) P_i(\mathbf{u}) P_j(\mathbf{u}_*) \, d\mathbf{u} \, d\mathbf{u}_* . \tag{2.5.11}$$

2.5.3 Models with Vanishing Space and Velocity Structure

The mathematical framework described in Subsection 2.5.1 corresponds to the spatially homogeneous case for particles in a suitable configuration (perhaps equilibrium), where the velocity distribution is constant in time.

On the other hand, some physical systems (an example is given in Chapter 5) are such that space and velocity are not significant in the model as the interacting particles are individuals which communicate at distances (psychologically, by media, etc.). The class of systems here technically differs from those analyzed in Subsection 2.5.1, where space and velocity variables are significant in describing the microscopic state of the active particles. The evolution equation (2.5.3) is obtained by averaging over space and velocity.

In the case here, the above averaging cannot be performed simply because $\mathbf{x}$ and $\mathbf{v}$ do not exist. However, one can heuristically derive the equation simply assuming that the encounter rate depends on the microscopic state. Therefore, equating the variation rate of the number of particles in the elementary volume $[\mathbf{u}, \mathbf{u} + d\mathbf{u}]$ to the net flux rate given by the interactions, but letting the interaction rate depend on the activities, yields

$$\frac{\partial}{\partial t} f_i^a(t, \mathbf{u}) = C_i^+[\mathbf{f}] - C_i^-[\mathbf{f}] + C_i[\mathbf{f}]$$

$$= \sum_{j=1}^{n} \int_{D_{\mathbf{u}} \times D_{\mathbf{u}}} \eta_{ij}(\mathbf{u}_*, \mathbf{u}^*) \mathcal{B}_{ij}(\mathbf{u}_*, \mathbf{u}^*; \mathbf{u}) f_i^a(t, \mathbf{u}_*) f_j^a(t, \mathbf{u}^*) \, d\mathbf{u}_* \, d\mathbf{u}^*$$

$$- f_i^a(t, \mathbf{u}) \sum_{j=1}^{n} \int_{D_\mathbf{u}} \eta_{ij}(\mathbf{u}, \mathbf{u}^*) f_j^a(t, \mathbf{u}^*) \, d\mathbf{u}^*$$

$$+ f_i^a(t, \mathbf{u}) \sum_{j=1}^{n} \int_{D_\mathbf{u}} \eta_{ij}(\mathbf{u}, \mathbf{u}^*) \mu_{ij}(\mathbf{u}, \mathbf{u}^*) f_j^a(t, \mathbf{u}^*) \, d\mathbf{u}^* . \qquad (2.5.12)$$

If the source/sink term is qual to zero, this structure corresponds to the Jager and Segel model, briefly described in Section 1.4, generalized to a system of several interacting populations.

A further case refers to systems where the space is a microscopic variable, but the velocity does not have a physical meaning in the model. This particular case will be examined in Chapter 4 which deals with systems where the microscopic activity state is a discrete variable rather than a continuous one.

2.6 Additional Concepts

Two conceivable mathematical frameworks, and some technical particularizations, have been developed in this chapter. The first structure corresponds to localized interactions, the second one to long range interactions. These two classes of equations should be regarded as a reference mathematical framework for the derivation of specific models related to specific physical systems, that is, referred to the modeling of microscopic interactions which differ from system to system. The main difficulty appears to be the modeling of the interplay between mechanical and activity quantities.

A large variety of models can be framed into these mathematical structures. However, alternative classes of equations can be developed, considering that specific applications can possibly suggest additional generalizations. This topic, which is an interesting and challenging research field for applied mathematicians, will be discussed later in this book. For instance, a variety of mathematical structures is reported in the interesting book by Schweitzer (2003) which deals with the derivation of master equations for active particles. It is a different approach which needs different paradigms, addressed to the derivation of evolution equations for the distribution function over the microscopic state of interacting particles.

The sequential steps to be used in deriving specific models consists in specializing the mathematical frameworks proposed in this chapter referring to the specific system to be modeled. The first step is the selection of the different populations which play the game and identification of the microscopic state of active particles within each population. Subsequently, one has to identify and model the different types of microscopic interactions which effectively occur: short and/or long range, conservative and/or proliferative or destructive. Implementing the models of microscopic interactions into the properly selected mathematical framework generates the model.

Different physical systems in the life sciences generally require different mathematical structures. Specific applications (models) may require some additional modifications of the mathematical structures (frameworks) developed in this chapter, when it is useful to deal with models where the microscopic state is a discrete, rather than continuous, variable.

Finally, computational methods provide a quantitative description of the evolution of the distribution function, while macroscopic quantities are obtained by weighted moments.

3

Additional Mathematical Structures for Modeling Complex Systems

---

3.1 Introduction

This chapter develops some generalizations of the mathematical structures proposed in Chapter 2, in view of research perspectives on advanced modeling. The models reported in Chapters 5–8 are derived based on the frameworks derived in Chapter 2 and their modifications for particles with discrete states, which are described in Chapter 4. Further modeling developments can possibly take advantage of the additional structures described in this chapter.

The various generalizations refer to particular frameworks rather than to the general case, for instance generalizations refer to models in the case of spatial homogeneity. The presentation is concise—the reader can develop detailed calculations following the guidelines of Chapter 2.

This chapter is organized into four more sections.

– Section 3.2 derives frameworks for particles undergoing mixed-type interactions, short and long range, that may be useful in modeling vehicular and pedestrian traffic phenomena, where interactions of the test active particle refer to all particles in its visibility zone. The interaction rules differ for close (almost colliding) and distant particles.

– Section 3.3 develops a mathematical structure for systems where interactions between active particles may possibly generate new particles in a population different from that of the interacting pair. This framework appears to be particularly useful for modeling complex systems in biological sciences, for instance, multicellular systems, where this type of event may take place due to genetic mutations.

– Section 3.4 outlines some alternatives to the modeling of microscopic interaction and space dynamics by using phenomenological models. This approach overcomes the difficulty of identifying a detailed description of microscopic interactions.

– Section 3.5 provides some additional reasoning focused on modeling the presence of external actions that may be related to control and optimization problems.

This chapter only covers some of the conceivable generalizations. Additional ones are proposed in the last chapter of this book.

3.2 Models with Mixed-Type Interactions

This section reports mathematical structures for modeling systems where microscopic interactions may be of mixed type, either short or long range. Encounters between particles belonging to certain pairs of populations follow short range interaction rules, while long range interactions occur with the other particles. The modeling is developed in the absence of external actions, leaving additional calculations to the initiative of the reader.

Let us denote by $A(j;i)$ the set of active particles of the jth population with short range interactions with active particles of the ith population, while $B(j;i)$ is the set related to long range interactions. Then $A(j;i) + B(j;i)$ represent, for each ith population, the whole set of interacting particles of the jth population.

Moreover, let us consider only the case of spatial homogeneity. The only mathematical structure, in the absence of external actions, is as follows:

$$\frac{\partial}{\partial t} f_i(t, \mathbf{u}) + \nabla_{\mathbf{u}} \cdot \left(\sum_{j \in B(j;i)} \mathcal{F}_{ij}^a[f^a] f_i^a \right)(t, \mathbf{u})$$

$$= \sum_{j \in A(j;i)} \eta_{ij}^0 \int_{D_{\mathbf{u}} \times D_{\mathbf{u}}} \mathcal{B}_{ij}(\mathbf{u}_*, \mathbf{u}^*; \mathbf{u}) f_i^a(t, \mathbf{u}_*) f_j^a(t, \mathbf{u}^*) \, d\mathbf{u}_* \, d\mathbf{u}^*$$

$$- f_i^a(t, \mathbf{u}) \sum_{j \in A(j;i)} \eta_{ij}^0 \int_{D_{\mathbf{u}}} f_j^a(t, \mathbf{u}^*)[1 - \mu_{ij}(\mathbf{u}, \mathbf{u}_*)] \, d\mathbf{u}^*$$

$$+ f_i^a(t, \mathbf{u}) \sum_{j \in B(j;i)} \int_{D_{\mathbf{u}}} \sigma_{ij}(\mathbf{u}, \mathbf{u}_*) f_j^a(t, \mathbf{u}_*) \, d\mathbf{u}_* , \qquad (3.2.1)$$

where

$$\mathcal{F}_{ij}^a = \int_{D_{\mathbf{u}}} \mathcal{P}_{ij}^a(\mathbf{u}, \mathbf{u}_*) f_j^a(t, \mathbf{u}_*) \, d\mathbf{u}_* . \qquad (3.2.2)$$

Mathematical frameworks with space structure are obtained using the full expression of the interaction terms given in Sections 2.3 and 2.4.

The above splitting can be precisely referred to the type of interactions. For instance, mechanical interactions can be assumed to be of long range type considering that active particles feel reciprocal presence even at long distances, while activity interactions are assumed to be short range due to binding phenomena between active particles which are possible only when a contact is realized. This structure can be useful (see Bellouquid and Delitala (2005)), for modeling cellular dynamics where the movement of cells is caused by long range interactions, while the biological dynamics is related to short range interactions.

3.3 Models with Exotic Proliferations

Various systems of interest in life sciences are characterized by a proliferative dynamics such that the interacting pairs may generate active particles in a population different from their original populations. This particular behavior may take place, for instance, in biological multiparticle systems when the interactions involve genetic mutations which shift particles from one population to another characterized by different genetic properties. See Hanahan and Weinberg (2000). These phenomena also occur in birth processes, in the animal world, when the interacting pair generates individuals with different features.

The analysis of this type of interactions is fully developed in the paper by De Lillo, Salvatori, and Bellomo (2007), where it is shown that these events can be generated by both conservative and proliferative interactions. The deductions proposed in that paper are reported here for the models with dominant activity interactions previously discussed in Section 2.5. The generalization to space-dependent problems requires nontrivial additional calculations that can be developed referring, when needed, to specific models.

Let us consider the case of localized interactions. Specifically, conservative interactions refer to the **candidate particle**, with state $\mathbf{u}_*$, of the hth population and the **field particle**, with state $\mathbf{u}^*$, of the kth population. Interactions occur with rate η_{hk}^0, while transitions are described by the term $\mathcal{B}_{hk}^i(\mathbf{u}_*, \mathbf{u}^*; \mathbf{u})$ which denotes the probability density that the candidate particle will fall into the state $\mathbf{u}$ of the ith population. It has to be regarded as a probability density with respect to the whole set of populations:

$$\forall\ \mathbf{u}_*, \mathbf{u}^*, \quad \forall\ h, k\ :\quad \sum_{i=1}^{n} \int_{D_{\mathbf{u}}} \mathcal{B}_{hk}^{i}(\mathbf{u}_*, \mathbf{u}^*; \mathbf{u})\, d\mathbf{u} = 1\,. \tag{3.3.1}$$

The flow G_i into the elementary volume of the state space, for the ith population, is given according to the above notation as follows:

$$G_i = \sum_{h=1}^{n} \sum_{k=1}^{n} \eta_{hk}^{0} \int_{D_{\mathbf{u}} \times D_{\mathbf{u}}} \mathcal{B}_{hk}^{i}(\mathbf{u}_*, \mathbf{u}^*; \mathbf{u}) f_h^{a}(t, \mathbf{u}_*) f_k^{a}(t, \mathbf{u}^*)\, d\mathbf{u}_*\, d\mathbf{u}^*\,, \tag{3.3.2}$$

while the loss term L_i remains the same.

In order to deal with nonconservative interactions, it is useful to split the term μ into the proliferative and destructive parts μ^p and μ^d, respectively. Destructive events occur in the same population of the interacting pairs, therefore the modeling is precisely the same as in Section 2.3. However, the dynamics of proliferative events follow rules analogous to those we have seen for conservative encounters. Therefore, the flux P_i^p in the space of elementary states of the ith population is given by

$$P_i^p = \sum_{h=1}^{n} \sum_{k=1}^{n} \eta_{hk}^{0} \int_{D_{\mathbf{u}} \times D_{\mathbf{u}}} \mu_{hk}^{i}(\mathbf{u}_*, \mathbf{u}^*; \mathbf{u}) f_h^{a}(t, \mathbf{u}_*) f_k^{a}(t, \mathbf{u}^*)\, d\mathbf{u}_*\, d\mathbf{u}^*\,.$$
$$\tag{3.3.3}$$

If interactions do not modify the state of the candidate particle, then the transition involves only test particles:

$$P_i^p = \sum_{h=1}^{n} \sum_{k=1}^{n} \eta_{ij}^{0} \int_{D_{\mathbf{u}}} \mu_{hk}^{i}(\mathbf{u}, \mathbf{u}^*) f_h^{a}(t, \mathbf{u}) f_k^{a}(t, \mathbf{u}^*)\, d\mathbf{u}^*\,. \tag{3.3.4}$$

The evolution equation finally results as follows:

$$\frac{\partial}{\partial t} f_i^{a}(t, \mathbf{u}) = \sum_{h=1}^{n} \sum_{k=1}^{n} \eta_{hk}^{0} \int_{D_{\mathbf{u}} \times D_{\mathbf{u}}} \mathcal{B}_{hk}^{i}(\mathbf{u}_*, \mathbf{u}^*; \mathbf{u}) f_h^{a}(t, \mathbf{u}_*) f_k^{a}(t, \mathbf{u}^*)\, d\mathbf{u}_*\, d\mathbf{u}^*$$

$$- f_i^{a}(t, \mathbf{u}) \sum_{j=1}^{n} \int_{D_{\mathbf{u}}} \eta_{ij}^{0} f_j^{a}(t, \mathbf{u}^*)\, d\mathbf{v}^*\, d\mathbf{u}^*$$

$$+ \sum_{h=1}^{n} \sum_{k=1}^{n} \eta_{hk}^{0} \int_{D_{\mathbf{u}} \times D_{\mathbf{u}}} \mu_{hk}^{i}(\mathbf{u}_*, \mathbf{u}^*; \mathbf{u}) f_h^{a}(t, \mathbf{u}_*) f_k^{a}(t, \mathbf{u}^*)\, d\mathbf{u}_*\, d\mathbf{u}^*$$

$$- f_i^{a}(t, \mathbf{u}) \sum_{j=1}^{n} \int_{D_{\mathbf{u}}} \eta_{ij}^{0} \mu_{ij}^{d}(\mathbf{u}, \mathbf{u}^*) f_j^{a}(t, \mathbf{u}^*)\, d\mathbf{v}^*\, d\mathbf{u}^*\,. \tag{3.3.5}$$

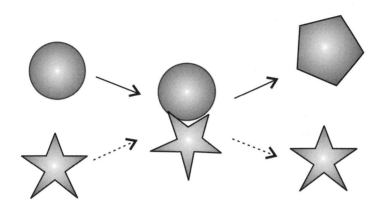

Fig. 3.3.1: Conservative exotic interactions.

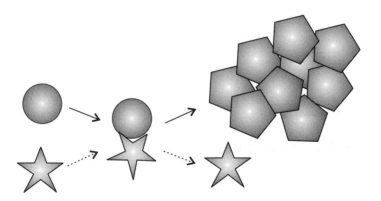

Fig. 3.3.2: Proliferative exotic interactions.

Figures 3.3.1 and 3.3.2 visualize the exotic proliferative dynamics.

The same reasoning can be generalized to long range interactions assuming that the proliferation terms σ can be split into the proliferative and destructive terms:

$$\sigma_{hk} = \sigma^p_{hk}(i) + \sigma^p_{hk}(h) - \sigma^d_{hk}, \qquad (3.3.6)$$

where the indexes in parentheses identify the population where the proliferation takes place, while it is understood that destruction occurs in the same population of the test particle.

The equation can be readily written for models with interactions involving predominantly the role of the activities:

$$\frac{\partial f_i^a}{\partial t}(t, \mathbf{u}) + \nabla_\mathbf{u} \cdot \left(\sum_{j=1}^{n} \mathcal{F}_{ij}^a[f^a] f_i^a \right)(t, \mathbf{u})$$

$$= \sum_{h=1}^{n} \sum_{k=1}^{n} \int_{D_\mathbf{u} \times D_\mathbf{u}} \sigma_{hk}^p(i)(\mathbf{u}, \mathbf{u}_*) \, f_h^a(t, \mathbf{u}^*) \, f_k^a(t, \mathbf{u}_*) \, d\mathbf{u}^* \, d\mathbf{u}_*,$$

$$+ f_i^a(t, \mathbf{u}) \sum_{j=1}^{n} \int_{D_\mathbf{u}} \sigma_{ij}^p(i)(\mathbf{u}, \mathbf{u}_*) \, f_j^a(t, \mathbf{u}_*) \, d\mathbf{u}_*$$

$$- f_i^a(t, \mathbf{u}) \sum_{j=1}^{n} \int_{D_\mathbf{u}} \sigma_{ij}^d(\mathbf{u}, \mathbf{u}_*) \, f_j^a(t, \mathbf{u}_*) \, d\mathbf{u}_* . \qquad (3.3.7)$$

The generalization to models with a space structure can be developed using the approach we have seen in the preceding chapter, while a technical simplification consists in modeling exotic proliferations by a linear additive term.

3.4 Phenomenological Frameworks

This section derives some phenomenological mathematical frameworks to be considered when the modeling of microscopic interactions cannot be mathematically described in all the details. The approach can be developed if the system shows a trend to an equilibrium function, say f_e. Then the evolution equation can be modeled according to the following formal framework:

$$\frac{\partial f}{\partial t} + \mathbf{v} \cdot \nabla_\mathbf{x} f = \mathcal{F}\big[f_e[f] - f\big], \qquad (3.4.1)$$

where $\mathcal{F}$ is a functional, acting over $f_e[f] - f$, suitable for describing the rate of trend to the equilibrium configuration $f_e[f]$, which may depend on f, for instance through its moments. Square brackets are used again to denote the object upon which the operator acts.

The BGK models of the kinetic theory of classical particles, see Cercignani, Illner, and Pulvirenti (1994), is technically analogous to Eq. (3.4.1). A quite general structure is as follows:

$$\frac{\partial f}{\partial t} + \mathbf{v} \cdot \nabla_\mathbf{x} f = \mathcal{C}[f_e]\big(f_e[f] - f\big), \qquad (3.4.2)$$

where the coefficient $\mathcal{C}$ may depend on f_e.

A simple model is obtained if the terms $\mathcal{C}$ and f_e are constant in time and uniform in space:

$$\frac{\partial f}{\partial t} + \mathbf{v} \cdot \nabla_{\mathbf{x}} f = c \left(f_{e0} - f \right), \tag{3.4.3}$$

where c is a constant and f_{e0} is a constant distribution. Moreover, only one of the two terms may be constant, while the other depends on f.

The term which expresses the trend to equilibrium can model an action additive with respect to the *term corresponding to localized or long-range interactions*. For instance, Eq. (2.5.3) can be written, for one population only, as follows:

$$\frac{\partial f}{\partial t} + \mathbf{v} \cdot \nabla_{\mathbf{x}} f = J[f] + \mathcal{C}[f]\big(f_e[f] - f\big), \tag{3.4.4}$$

where the term J identifies the right-hand side of Eq. (2.5.3).

The same reasoning can be applied to Eq. (2.5.5) as follows:

$$\frac{\partial f}{\partial t} + \mathbf{v} \cdot \nabla_{\mathbf{x}} f + T^m[\mathbf{f}] + T^a[f] = \mathcal{S}[f] + \mathcal{C}[f]\big(f_e[f] - f\big). \tag{3.4.5}$$

Observe that all these mathematical frameworks are derived assuming that active particles can freely move in space. However, this is not always the case, because local space obstacles may prevent some movements or identify preferred movement directions. Discrete velocity models can be developed to take this aspect into account if the velocity directions are related to the geometry of the environment.

A totally different alternative to these approaches consists in modeling the space dynamics by random walk models. A specific model that takes the above considerations into account is the following:

$$\frac{\partial f}{\partial t}(t, \mathbf{x}, \mathbf{v}, \mathbf{u}) + \mathbf{v} \cdot \nabla_{\mathbf{x}} f(t, \mathbf{x}, \mathbf{v}, \mathbf{u})$$

$$= \nu \int_{D_{\mathbf{v}}} \Big[T(\mathbf{v}, \mathbf{v}^*) f(t, \mathbf{x}, \mathbf{v}^*, \mathbf{u}) - T(\mathbf{v}^*, \mathbf{v}) f(t, \mathbf{x}, \mathbf{v}, \mathbf{u}) \Big] d\mathbf{v}^*$$

$$+ \eta^0 \Big[\int_{D_{\mathbf{u}}} \int_{D_{\mathbf{u}}} \mathcal{B}(\mathbf{u}_*, \mathbf{u}^*, \mathbf{u}) f(t, \mathbf{x}, \mathbf{v}, \mathbf{u}_*) f(t, \mathbf{x}, \mathbf{v}, \mathbf{u}^*) \, d\mathbf{u}_* \, d\mathbf{u}^*$$

$$- f(t, \mathbf{x}, \mathbf{v}, \mathbf{u}) \int_{D_{\mathbf{u}}} [1 - \mu(\mathbf{u}, \mathbf{u}^*)] f(t, \mathbf{x}, \mathbf{v}, \mathbf{u}^*) d\mathbf{u}^* \Big], \tag{3.4.6}$$

where ν is the turning rate or turning frequency (hence $\tau = \frac{1}{\nu}$ is the mean run time); and $T(\mathbf{v}, \mathbf{v}^*)$ is the probability kernel over the new velocity $\mathbf{v} \in D_{\mathbf{v}}$ if the previous velocity was $\mathbf{v}^*$.

The above notation corresponds to the assumption that active particles choose any direction with bounded velocity. Specifically, the set of possible velocities is denoted by $D_{\mathbf{v}}$, where $D_{\mathbf{v}} \subset \mathbb{R}^3$, and it is assumed that $D_{\mathbf{v}}$ is bounded and symmetric (i.e., $\mathbf{v} \in D_{\mathbf{v}} \Rightarrow -\mathbf{v} \in D_{\mathbf{v}}$). Therefore, the dynamics is modeled by the probability kernel T, while the evolution related to the activity is ruled by predominant activity interactions.

This type of modeling appears to be of interest in biological sciences and it has been applied in the paper by Bellomo and Bellouquid (2006) to derive macroscopic equations corresponding to an underlying microscopic description. It has been shown how different macroscopic models can be related to different models of microscopic interactions. This result can be regarded as an additional indication of the complex interplay between mechanical microscopic variables and activity, that can substantially modify the overall behavior of the system.

This reasoning can also be applied to interactions ruled by long range actions. In this case, the evolution equation is

$$
\frac{\partial f}{\partial t}(t, \mathbf{x}, \mathbf{v}, \mathbf{u}) + \mathbf{v} \cdot \nabla_{\mathbf{x}} f(t, \mathbf{x}, \mathbf{v}, \mathbf{u}) + \nabla_{\mathbf{u}} \cdot (\mathcal{F}[f]f)(t, \mathbf{x}, \mathbf{v}, \mathbf{u})
$$

$$
= \nu \int_{D_{\mathbf{v}}} \left[T(\mathbf{v}, \mathbf{v}^*)f(t, \mathbf{x}, \mathbf{v}^*, \mathbf{u}) - T(\mathbf{v}^*, \mathbf{v})f(t, \mathbf{x}, \mathbf{v}, \mathbf{u}) \right] d\mathbf{v}^*
$$

$$
+ f(t, \mathbf{x}, \mathbf{v}, \mathbf{u}) \int_{D_{\mathbf{u}}} \sigma_{ij}(\mathbf{u}, \mathbf{u}_*) \, f_j(t, \mathbf{x}, \mathbf{v}, \mathbf{u}_*) \, d\mathbf{u}_* , \qquad (3.4.7)
$$

where

$$
\mathcal{F} = \int_{D_{\mathbf{u}}} \mathcal{P}(\mathbf{u}, \mathbf{u}_*) \, f(t, \mathbf{x}, \mathbf{v}, \mathbf{u}_*) \, d\mathbf{u}_* . \qquad (3.4.8)
$$

More generally, the turning operator may depend on both variables: velocity and activity. Hence, this term is written as follows:

$$
\nu \int_{D_{\mathbf{v}} \times D_{\mathbf{u}}} \left[T(\mathbf{v}, \mathbf{u}, \mathbf{v}^*, \mathbf{u}^*)f(t, \mathbf{x}, \mathbf{v}^*, \mathbf{u}^*) \right.
$$

$$
\left. - T(\mathbf{v}^*, \mathbf{u}^*, \mathbf{v}, \mathbf{u})f(t, \mathbf{x}, \mathbf{v}, \mathbf{u}) \right] d\mathbf{u}_* \, d\mathbf{v}_* , \qquad (3.4.9)
$$

to be substituted into Eqs. (3.4.6) or (3.4.7).

Looking at additional conceivable frameworks, one can also consider models where the turning operator modifies the activities, while the dynamics to related mechanical activities is described by phenomenological models like those we have seen above. These must be regarded as simple suggestions to be properly developed for specific applications.

3.5 Open Systems

A large variety of models can be framed into the mathematical structures described until now, but alternative classes of equations can also be developed, possibly suggested by specific applications. This topic, which is an interesting and challenging research field for applied mathematicians, will be further analyzed later in this book. However, the derivation of the mathematical structures considered until now, refers to closed systems in the absence of external actions. These actions can be applied to modify the distribution function even in the absence of microscopic interactions among active particles.

The modeling of these external actions is outlined in the relatively simple case of systems with dominant activity interactions, i.e., for the structures underlying the equations modeling the evolution of the activity variable for spatial homogeneity and short range interactions. The extension to more general cases requires additional technical calculations and is left to the reader possibly in connection with specific applications.

Let us consider the case of one population only of particles under an external action which modifies their scalar microscopic state:

$$\frac{du}{dt} = k(t, u).$$
(3.5.1)

The action $k(t, u)$ acts as a transport term. Consequently, the mathematical structure is as follows:

$$\frac{\partial}{\partial t} f^a(t, u) + \frac{\partial}{\partial u}\big(k(t, u) f^a(t, u)\big) = J[f^a](t, u),$$
(3.5.2)

where $J[f^a]$ is the right-hand term of the equation corresponding to particle interactions.

Linearity of the transport term appears if the action depends on time only:

$$\frac{\partial}{\partial t} f^a(t, \mathbf{u}) + k(t) \frac{\partial}{\partial u} f^a(t, u) = J[f^a](t, u).$$
(3.5.3)

This equation includes a deterministic action in the left-hand side term. An alternative consists in introducing a stochastic modeling of the presence of the external actions at the microscopic level by a known distribution $g(t, u)$. The framework corresponding to short range interactions is as follows:

$$\frac{\partial}{\partial t} f^a(t, u) = J[f^a](t, u)$$

$$+ \int_{D_u \times D_u} \eta^*(u_*, u^*) \mathcal{C}(u_*, u^*; u) f^a(t, u_*) g(t, u^*) \, du_* \, du^*$$

$$- f^a(t, u) \int_{D_u} \eta^*(u, u^*) g(t, u^*) \, du^*$$

$$+ f^a(t, u) \int_{D_u} \eta^*(u, u^*) \mu(u, u^*) g(t, u^*) \, du^* , \qquad (3.5.4)$$

where η^* is the encounter rate between active particles and the external action at the microscopic level. In general η^* can be scaled with respect to η, namely $\eta^* = \varepsilon \eta$.

Of course, the same modeling can be straightforwardly applied to long range interactions, as well as to models depending on space. Detailed calculations are not reported here. A deeper insight can be referred to specific applications.

4

Mathematical Frameworks
for Discrete Activity Systems

4.1 Introduction

Chapters 2 and 3 were devoted to the analysis of models where the microscopic state, specifically, the *activity* was assumed to be a continuous variable defined over bounded or unbounded domains. On the other hand, a variety of physical systems in the life sciences are characterized by active particles which need a discrete variable to describe their microscopic state. Therefore, it is useful to format the mathematical tools derived in the preceding chapters to this specific case.

Various applications proposed in the chapters which follow show how the modeling of the same system can be developed using either discrete or continuous variables. When both types of mathematical description of the microscopic state can be used, a critical analysis is developed to compare the two technically different ways of deriving the mathematical models.

Discrete activity models are not introduced to reduce complexity, either concerning modeling or computational aspects or both, but are motivated by specific modeling requirements. Namely, a discrete variable can be, in some cases, the correct way to describe the microscopic state.

The discretization of the microscopic state is generally referred to the activity. However, for some particular systems, it may be useful to deal also with the discretization of the velocity variable in a way technically analogous to the methods used to derive the discrete Boltzmann equation, which was briefly introduced in Chapter 1. Moreover, one should also consider the case where active particles may be allowed to move along particular selected directions.

This type of discretization is also useful when linked to specific features of the real system, and not used simply as a way to reduce the computational complexity related to models with a full range of velocities. Indeed, the motion of active particles is often constrained by upper and lower bounds, while in the case of the Boltzmann equations it is assumed that particles can attain all velocities in a three-dimensional space. Therefore, the computational complexity is not as relevant as it is for models of classical kinetic theory.

The same reasoning can also be applied to the space variable when active particles have the ability to shift from one zone to another by jumps rather than through a continuous movement. Relatively more detailed examples are given later in the chapter.

Therefore, it is necessary to develop a mathematical framework, analogous to that of Chapters 2 and 3, for models for which the microscopic activity is a discrete variable. The analysis is developed when the activity is discrete, and the space and velocity variables are left to be continuous.

This chapter consists of five more sections as follows.

– Section 4.2 analyzes some motivations and examples where the use of discrete variables is consistent with the physics of the system. A simple qualitative description of various physical systems illustrates this issue.

– Section 4.3 shows how some physical systems can be described by a discrete distribution function related to each population, and how macroscopic observable quantities can be recovered by averaged moments. The distribution function is left continuous over the space and velocity variables, while it is discrete over the activity variable. In addition, some technical generalizations to the case of distribution functions discrete on the whole microscopic variable are briefly analyzed.

– Section 4.4 derives a mathematical framework analogous to that developed in Chapter 2. The derivation is not a technical generalization of the previous one, as introducing discrete models requires additional analyses, while it also offers new mathematical tools to be used in the applications. Some of these tools play a relevant role in the modeling of living systems.

– Section 4.5 proposes various technical generalizations. First, it discusses systems where interactions between active particles of two different (or of the same) populations may possibly generate active particles in a third population. Subsequently, suitable frameworks obtained by discretization of the other microscopic, geometrical and mechanical, variables are developed.

– Section 4.6 provides a critical analysis concerning the effective descriptive ability of the mathematical tools proposed in this chapter. The analysis looks at some of the applications dealt with in the chapters which follow.

The contents of this chapter essentially refer to the paper by Bertotti and Delitala (2004), where the derivation of discrete generalized models was initiated. Further developments have been proposed by Chauviere and Brazzoli (2006) for closed systems, further developed by Brazzoli (2007) for open systems.

The dynamics of microscopic interactions is precisely the same as the one already analyzed in Chapter 2; therefore the reader is referred to the visualization offered by the figures of that chapter.

4.2 Motivations for a Discrete States Modeling

Modeling the microscopic activity of living particles by a discrete variable rather than a continuous one is motivated, as already mentioned, by modeling requirements rather than by the aim to reduce computational complexity. The assessment of the ***activity*** microscopic variable has to be precisely related to the particular physical system to be mathematically described. The strategy for the identification of the microscopic variable may vary from system to system also depending on its observability and on the effective possibility of developing experimental measurements. Some examples of models will be analyzed in what follows. The strategy for the discretization of the other microscopic variables, position and/or velocity, is also considered.

Let us consider, as an example, the class of physical systems in Chapter 5 related to the modeling of various aspects of the social competition among individuals characterized by a microscopic state referred to their social collocation. It is plain that it is technically reasonable to model the social state by a discrete variable related to various levels of wealth (or poverty), rather than to consider the microscopic variable as a continuous one. Namely, social levels, in some cases social classes, can be practically identified only within intervals corresponding to discrete states. The same reasoning applies to additional variables, for instance, the level of education, which may also be used to describe the state of social systems.

The microscopic variable suitable to describe the active particles (individuals) within this system may also be discrete with reference to the space, so that living areas can be identified, while the velocity variable does not play, in this case, a relevant role. This aspect suggests that we develop an additional analysis for which the whole microscopic state is described by discrete variables.

Chapter 6 deals with vehicular traffic flow modeling. If drivers are considered as active particles, the driver-vehicles can be subdivided into various categories: trucks, slow cars driven by beginners, vehicles driven by expert drivers, very fast cars possibly driven by crazy drivers, and so on.

In this case, the space variable is continuous, while the activity variable may be discrete according to the above classification of vehicle categories. Moreover, it is shown that the velocity variable can also be assumed to be discrete to model the granular (rather than continuous) feature of the flow. Discrete velocity intervals capture groups of vehicles without the assumption of a continuous velocity distribution, which is reasonable only for a large number of interacting particles as in fluid dynamics.

The same reasoning can be applied to Chapter 7, which deals with multicellular biological systems. Here the activity refers to biological functions which may be referred to genetic mutations that induce a sharp, rather than continuous, variation of activity. See Vogelstein and Kinzler (2004). Therefore, we see that a discrete representation can be, in some cases, more consistent with physical reality. In some cases, both modeling approaches: continuous and discrete, are valid.

4.3 On the Discrete Distribution Function

Let us consider a system of active particles organized in n interacting populations labeled by the subscripts $i = 1, \ldots, n$, where each population is characterized by a different way of organizing their specific activities as well as their interactions with the other populations. Accordingly, the activity of the particles can be represented by the set

$$I_{\mathbf{u}} = \{\mathbf{u}_1, \ldots, \mathbf{u}_h, \ldots, \mathbf{u}_H\}, \qquad (4.3.1)$$

with components $\mathbf{u}_h$, where $h = 1, \ldots, H$.

The statistical distribution suitable for describing the state of the system is the discrete one-particle distribution function

$$f_i^h = f_i^h(t, \mathbf{x}, \mathbf{v}) = f_i(t, \mathbf{x}, \mathbf{v}; \mathbf{u}_h), \quad [0, T] \times D_{\mathbf{x}} \times D_{\mathbf{v}} \to \mathbb{R}_+, \quad (4.3.2)$$

corresponding to the ith population and to the hth state, where the space and velocity variables are still assumed to be continuous.

The set of all distributions is denoted by

$$f_i = \{f_i^h\}_{h=1}^{h=H}, \qquad \text{and} \qquad \mathbf{f} = \{f_i\}_{i=1}^{i=n}.$$

Similarly to the case of continuous distribution functions, the knowledge of f_i^h leads to the computation of the macroscopic quantities. Calculations are analogous to those in Chapter 2: simply integration is replaced by sums. Therefore, computations are proposed only for some of the macroscopic variables to be taken as examples. The reader has already received sufficient information to complete the analysis.

The **number density**, or **size**, of active particles at time t in position **x** for the ith population is

$$n_i[f_i](t, \mathbf{x}) = \sum_{h=1}^{H} \int_{D_\mathbf{v}} f_i^h(t, \mathbf{x}, \mathbf{v}) \, d\mathbf{v} \, . \tag{4.3.3}$$

Moreover, the total density is obtained by summing the densities corresponding to all populations:

$$n[\mathbf{f}](t, \mathbf{x}) = \sum_{i=1}^{n} n_i[f_i](t, \mathbf{x}) \, , \tag{4.3.4}$$

while the initial sizes are denoted by

$$n_{i0}(\mathbf{x}) = n_i(t = 0, \mathbf{x}) \, , \qquad \text{and} \qquad n_0(\mathbf{x}) = n(t = 0, \mathbf{x}) \, .$$

Integration over the volume $D_\mathbf{x}$ containing the particles gives the **total size** of the ith population: $N_i(t)$. The total size $N = N(t)$ of all populations is given by the sum of all N_i. In this case, the densities f_i^h can be normalized, dividing them by $N_0 = N(t = 0)$.

First-order moments provide either **linear mechanical macroscopic** quantities, or **linear activity macroscopic** quantities. For instance, the mass velocity of particles, at time t in position **x**, is defined by

$$\mathbf{U}[f_i](t, \mathbf{x}) = \frac{1}{n_i(t, \mathbf{x})} \sum_{h=1}^{H} \int_{D_\mathbf{v}} \mathbf{v} \, f_i^h(t, \mathbf{x}, \mathbf{v}) \, d\mathbf{v} \, . \tag{4.3.5}$$

Moreover, focusing on activity terms, linear moments related to each hth component of the state **u**, related to the ith population, can be called hth **activation** at time t in position **x**, and are computed as follows:

$$\mathbf{a}_i^h = \mathbf{a}_i^h[f_i^h](t, \mathbf{x}) = \mathbf{u}_h \int_{D_\mathbf{v}} f_i^h(t, \mathbf{x}, \mathbf{v}) \, d\mathbf{v} \, . \tag{4.3.6}$$

The hth **activation density** is given by

$$\mathbf{d}_i^h = \mathbf{d}_i^h[f_i](t, \mathbf{x}) = \frac{\mathbf{a}_i^h[f_i^h](t, \mathbf{x})}{n_i[f_i](t, \mathbf{x})} \, . \tag{4.3.7}$$

Global quantities are obtained by integrating over space. The **global activation** is

$$\mathbf{A}_i^h = \mathbf{A}_i^h[f_i^h](t) = \int_{D_{\mathbf{x}}} \mathbf{a}_i^h(t, \mathbf{x}) \, d\mathbf{x}, \tag{4.3.8}$$

while the **global activation density** is given by

$$\mathbf{D}_i^h = \mathbf{D}_i^h[f_i](t) = \int_{D_{\mathbf{x}}} \mathbf{d}_i^h(t, \mathbf{x}) \, d\mathbf{x}. \tag{4.3.9}$$

Similar calculations can be developed for higher-order moments if motivated by some interest for the application. For instance, the hth **quadratic activity**, corresponding to the j-component u_{hj} of $\mathbf{u}_h$, can be computed as second-order moments:

$$e_{ij}^h = e_{ij}^h[f_i^h](t, \mathbf{x}) = u_{hj}^2 \int_{D_{\mathbf{v}}} f_i^h(t, \mathbf{x}, \mathbf{v}) \, d\mathbf{v}, \tag{4.3.10}$$

while the hth **quadratic activity density** is given by

$$\varepsilon_{ij}^h = \varepsilon_{ij}^h[f_i](t, \mathbf{x}) = \frac{e_{ij}^h[f_i^h](t, \mathbf{x})}{n_i[f_i](t, \mathbf{x})}. \tag{4.3.11}$$

Summing over h provides quantities for the whole population. For instance, linear and quadratic activations are, respectively, given by

$$\mathbf{a}_i = \sum_{h=1}^{H} \mathbf{a}_i^h[f_i^h](t, \mathbf{x}), \tag{4.3.12}$$

and

$$e_i = \sum_{h=1}^{H} e_i^h[f_i^h](t, \mathbf{x}). \tag{4.3.13}$$

Analogous calculations can be made for the other quantities, and again, global quantities are obtained by integrating over the space variable.

The above reasoning can be easily generalized to systems of active particles when the other microscopic variables are also discrete. For instance, both space and velocity variables may be discrete due to modeling requirements. Referring to the models outlined in Section 4.2, social dynamics may require the identification of a limited number of areas where individuals live, selecting them according to their social state. Traffic flow modeling, as already mentioned, requires the assessment of discrete velocity variables related to different types of driver-vehicle systems, e.g., slow and fast vehicles.

If the velocity variable is also discrete, belonging to the set

$$I_{\mathbf{v}} = \{\mathbf{v}_1, \dots, \mathbf{v}_k, \dots, \mathbf{v}_K\}, \qquad (4.3.14)$$

with components $\mathbf{v}_k$, $k = 1, \dots, K$, the distribution function is as follows:

$$f_i(t, \mathbf{x}, \mathbf{v}, \mathbf{u}) = \sum_{k=1}^{K} \sum_{h=1}^{H} f_i^{kh}(t, \mathbf{x})\, \delta(\mathbf{v} - \mathbf{v}_k)\, \delta(\mathbf{u} - \mathbf{u}_h), \qquad (4.3.15)$$

where the space variable remains continuous, and

$$f_i^{kh}(t, \mathbf{x}) = f_i(t, \mathbf{x}, \mathbf{v}_k, \mathbf{u}_h). \qquad (4.3.16)$$

Moments are still obtained according to the method we have seen above, simply replacing integration over the velocity variable by finite sums. For instance, the local density is given by

$$n_i[f_i](t, \mathbf{x}) = \sum_{k=1}^{K} \sum_{h=1}^{H} f_i^{kh}(t, \mathbf{x}), \qquad (4.3.17)$$

where $f_i = \{f_i^{hk}\}$.

Analogous calculations can be developed for first- and higher-order moments simply by replacing integration over the velocity variable by finite sums. When the space variable is also discrete, one has

$$I_{\mathbf{x}} = \{\mathbf{x}_1, \dots, \mathbf{x}_\ell, \dots, \mathbf{x}_L\}, \qquad (4.3.18)$$

so that the distribution function is as follows:

$$f_i(t, \mathbf{x}, \mathbf{v}, \mathbf{u}) = \sum_{\ell=1}^{L} \sum_{k=1}^{K} \sum_{h=1}^{H} f_i^{\ell k h}(t)\, \delta(\mathbf{x} - \mathbf{x}_\ell)\, \delta(\mathbf{v} - \mathbf{v}_k)\, \delta(\mathbf{u} - \mathbf{u}_h), \quad (4.3.19)$$

where

$$f_i^{\ell k h}(t) = f_i(t, \mathbf{x}_\ell, \mathbf{v}_k, \mathbf{u}_h), \qquad (4.3.20)$$

and where $f_i = \{f_i^{h\ell k}\}$.

Moment calculations are analogous to those for the distribution function with discrete activity variable only.

4.4 Mathematical Framework

This section derives a general mathematical framework for the evolution of the above discrete distribution functions f_i^h, analogous to the one we have seen for continuous distribution functions. First, the spatially homogeneous case is considered, followed by a generalization to systems characterized by space dynamics.

The approach is technically different from that in Chapter 2, and it is useful to show how the same result can be obtained following a different reasoning. The derivation of mathematical frameworks here refers to localized interactions; long range interactions need smooth evolutions which may not always be consistent with discrete variables. These topics are discussed in the following subsections.

4.4.1 Microscopic State Identified by the Activity Only

Let us first consider the simplest type of discretization which refers to the case of spatial homogeneity, namely when the distribution function depends on the activity only. In this case, the distribution function is written as a sum of Dirac distributions:

$$f_i(t, \mathbf{u}) = \sum_{h=1}^{H} f_i^h(t)\, \delta(\mathbf{u} - \mathbf{u}_h)\,, \qquad (4.4.1)$$

where, for simplicity of notation, f replaces f^a.

The active particles are then homogeneously distributed in space undergoing **localized binary interactions** involving **test** or **candidate** particles and **field** particles, in which the test particle enters into the action domain of the field particle. The action domain of the field particle is relatively small so that only binary encounters are relevant.

Following the method described in Chapter 2, let us consider the following classification of interactions:

• **Conservative interactions** which modify the microscopic state, mechanical and/or activity, of the interacting active particles, but not the size of the population;

• **Proliferative** or **destructive interactions** with death or birth of active particles due to pair interactions.

The modeling of microscopic interactions is based on the assumption that the following three quantities can be computed.

• The **interaction rate**. This is assumed to depend only on the pair of interacting populations i and j and not on the activities of the interacting particles: $\eta_{ij} = c_{ij}$.

- The *transition probability density*:

$$\mathcal{B}_{ij}^{pq}(h) = \mathcal{B}_{ij}(\mathbf{u}_p, \mathbf{u}_q; \mathbf{u}_h),\tag{4.4.2}$$

which is the probability density that a **candidate** particle with state $\mathbf{u}_p$, of the ith population will fall into the state $\mathbf{u}_h$ of the **test** particle (of the same population) after an interaction with a **field** particle of the jth population with state $\mathbf{u}_q$.

The transition density function (4.4.2) has the structure of a probability density with respect to the variable $\mathbf{u}_h$, namely, with respect to the index h:

$$\forall i,j \quad \forall p,q: \quad \sum_{h=1}^{H} \mathcal{B}_{ij}^{pq}(h) = 1.\tag{4.4.3}$$

- The **proliferative/destructive term** $\mu_{ij}^{pq}(\mathbf{u}_p, \mathbf{u}_q)$, where $\eta_{ij}\,\mu_{ij}^{pq}$ is the self-proliferation or self-destruction rate of an active particle with state $\mathbf{u}_p$ due to its interactions, occurring with the above-defined **interaction rate**, with the **field** particle with state $\mathbf{u}_q$. Interactions occur between particles of the populations i and j with proliferation or destruction in the same population of the test particle.

The representation of the preceding interactions is technically analogous to those we have seen in Chapter 2. The difference is that they depend on the discrete activity.

The balance of particles in the space of microscopic states generates the following class of equations:

$$\frac{df_i^h}{dt} = \sum_{j=1}^{n} \left(\sum_{p=1}^{H}\sum_{q=1}^{H} c_{ij}\mathcal{B}_{ij}^{pq}(h)f_i^p f_j^q - f_i^h \sum_{q=1}^{H} c_{ij}\, f_j^q + f_i^h \sum_{q=1}^{H} c_{ij}\mu_{ij}^{hq}\, f_j^q \right),$$
$$\tag{4.4.4}$$

for $i = 1,\ldots,n$, and $h = 1,\ldots,H$, and where $f_i^h = f_i^h(t)$.

Equation (4.4.4) defines a system of $n \times H$ ordinary differential equations that, for conservative interactions only, is written as follows:

$$\frac{df_i^h}{dt} = \sum_{j=1}^{n} \left(\sum_{p=1}^{H}\sum_{q=1}^{H} c_{ij}\mathcal{B}_{ij}^{pq}(h)f_i^p f_j^q - f_i^h \sum_{q=1}^{H} c_{ij}f_j^q \right).\tag{4.4.5}$$

In the case of one population only it simplifies into the following system of H ordinary differential equations:

$$\frac{df^h}{dt} = \sum_{p=1}^{H}\sum_{q=1}^{H} c\mathcal{B}^{pq}(h)f^p f^q - f^h \sum_{q=1}^{H} c(1 - \mu^{pq})f^q.\tag{4.4.6}$$

Similarly to Chapter 2, it is necessary to distinguish the above case of spatial homogeneity, where the velocity distribution has reached a constant distribution, from the case of systems where the microscopic state effectively depends on the activity variable only, and where the velocity variable does not has a physical meaning.

In the latter case, the encounter rate may also depend on the microscopic state of the interacting pair, and Eqs. (4.4.5) and (4.4.6) have to be rewritten as follows:

$$\frac{df_i^h}{dt} = \sum_{j=1}^{n} \left(\sum_{p=1}^{H} \sum_{q=1}^{H} c_{ij}^{pq} \mathcal{B}_{ij}^{pq}(h) f_i^p f_j^q - f_i^h \sum_{q=1}^{H} c_{ij}^{hq} f_j^q \right). \qquad (4.4.7)$$

in the case of n populations, and

$$\frac{df^h}{dt} = \sum_{p=1}^{H} \sum_{q=1}^{H} c^{pq} \mathcal{B}^{pq}(h) f^p f^q - f^h \sum_{q=1}^{H} c^{hq}(1 - \mu^{pq}) f^q, \qquad (4.4.8)$$

in the case of one population only.

4.4.2 Microscopic State also Dependent on Space and Velocity Variables

Let us now consider the case of models in which the space structure cannot be neglected, so that the discrete distribution function also depends on the space and velocity variables: $f_i^h(t, \mathbf{x}, \mathbf{v}) = f_i(t, \mathbf{x}, \mathbf{v}; \mathbf{u}_h)$.

In this case, the encounter rate can be thought of as depending on the relative velocity of the interacting pair as stated in Section 2.3. Moreover, similar to the case of spatial homogeneity, the transition probability density involving the velocity variable:

$$\mathcal{C}_{ij}^{pq}(h) = \mathcal{C}_{ij}(\mathbf{v}_1, \mathbf{u}_p, \mathbf{v}_2, \mathbf{u}_q; \mathbf{u}_h, \mathbf{v}), \qquad (4.4.9)$$

defines the probability density that an active particle with state $(\mathbf{v}_1, \mathbf{u}_p)$, belonging to the ith population, will fall into the state $(\mathbf{v}, \mathbf{u}_h)$ of the same population, after interaction with an active particle with state $(\mathbf{v}_2, \mathbf{u}_q)$, belonging to the jth population.

The above probability density can be assumed to be given as the product of the probability densities related to independent interactions of the activity and mechanical variables:

$$\mathcal{C}_{ij}^{pq}(h)(\cdot) = \mathcal{B}_{ij}^{pq}(h) \, \mathcal{M}_{ij}^{pq}(\mathbf{v}_1, \mathbf{v}_2; \mathbf{v} | \mathbf{u}_p, \mathbf{u}_q). \qquad (4.4.10)$$

Proliferation occurs with encounter rate η_{ij} and proliferation rate μ_{ij}^{pq}. Namely, interactions modify microscopic activity with rate η_{ij}, but with an output which depends only on the input activities of the interacting pair. On the other hand, the output of the velocities also depends on the input activities.

Therefore, applying the usual balance equation yields

$$\frac{\partial f_i^h}{\partial t}(t, \mathbf{x}, \mathbf{v}) + \mathbf{v} \cdot \nabla_{\mathbf{x}} f_i^h(t, \mathbf{x}, \mathbf{v}) = J_i^h[\mathbf{f}](t, \mathbf{x}, \mathbf{v}) = \sum_{j=1}^{n} J_{ij}^h[\mathbf{f}](t, \mathbf{x}, \mathbf{v}),$$

where (4.4.11)

$$J_{ij}^h = \sum_{p=1}^{H} \sum_{q=1}^{K} \int_{D_{\mathbf{v}} \times D_{\mathbf{v}}} c_{ij} |\mathbf{v}_1 - \mathbf{v}_2| \mathcal{B}_{ij}^{pq}(h)$$

$$\times \mathcal{M}_{ij}^{pq}(\mathbf{v}_1, \mathbf{v}_2; \mathbf{v} | \mathbf{u}_p, \mathbf{u}_q) f_i^p(t, \mathbf{x}, \mathbf{v}_1) f_j^q(t, \mathbf{x}, \mathbf{v}_2) \, d\mathbf{v}_1 \, d\mathbf{v}_2$$

$$- f_i^h(t, \mathbf{x}, \mathbf{v}) \sum_{q=1}^{K} \int_{D_{\mathbf{v}}} c_{ij} |\mathbf{v} - \mathbf{v}_2| f_j^q(t, \mathbf{x}, \mathbf{v}_2) \, d\mathbf{v}_2$$

$$+ f_i^h(t, \mathbf{x}, \mathbf{v}) \sum_{q=1}^{K} \int_{D_{\mathbf{v}}} c_{ij} |\mathbf{v} - \mathbf{v}_2| \mu_{ij}^{hq} f_j^q(t, \mathbf{x}, \mathbf{v}_2) \, d\mathbf{v}_2, \quad (4.4.12)$$

for $i = 1, \ldots, n$ and $h = 1, \ldots, H$.

If the interaction related to the mechanical variable is not affected by the activity variable and preserves (as seen in Chapter 1 for the discrete Boltzmann equations) momentum and energy, the following relations:

$$\mathbf{v}_1 + \mathbf{v}_2 = \mathbf{v}_1^* + \mathbf{v}_2^*,$$

$$v_1^2 + v_2^2 = v_1^{*2} + v_2^{*2}$$

hold, where the asterisks denote the velocities of the particles after the interaction.

This modeling assumes that active particles move along straight trajectories until another particle is encountered. However, the trajectory can be modified by the presence of the field applied by the other particles. This phenomenon can be modeled by long range interactions using the methods described in Chapter 2. For instance, it is possible to deal with systems which show a trend to an equilibrium state:

$$\frac{\partial f_i^h}{\partial t}(t, \mathbf{x}, \mathbf{v}) + \mathbf{v} \cdot \nabla_{\mathbf{x}} f_i^h(t, \mathbf{x}, \mathbf{v}) = \sum_{j=1}^{n} J_{ij}^h[\mathbf{f}](t, \mathbf{x}, \mathbf{v})$$

$$+ \varepsilon_i^h[f_i^{he}(\mathbf{x}, \mathbf{v}) - f_i^h(t, \mathbf{x}, \mathbf{v})], \quad (4.4.13)$$

where f_i^{he} denotes the equilibrium state corresponding to the component h of the $i^3 th$ population. More in general, ε_i^h and f_i^{he} may also depend on the distribution function.

4.5 Additional Generalizations

Some additional generalizations of the framework given in Section 4.4 are developed here in view of some specific applications. Specifically, the following cases are considered:

i) Systems where active particles can generate particles in a population different from the ones of the interacting pairs;

ii) Systems of active particles whose microscopic state is discrete, not only for the activity variable, but also in the space and/or velocity variables.

iii) Open systems of active particles subject to external actions.

These generalizations will receive a deeper analysis in the forthcoming chapters with reference to specific models.

The first generalization can be useful for applications in the modeling of biological systems of interacting cells where cell interactions and signaling may shift cells from one population to another due to genetic mutations, Delitala and Forni (2007).

The second generalization can be useful for applications in the modeling of traffic flow phenomena, when experimental measurements show that the identification of vehicles is technically possible only within fixed ranges of velocities. The number of vehicles is not large enough, unlike the case of gas particles, to justify the assumption of a continuous velocity distribution.

Finally, the third generalization attempts to model how a system of active particles interacts with the outer environment and it specifically focuses on the actions addressed to control the system or optimize its behavior.

In principle all systems analyzed in the second part of the book may be subject to external actions; for instance, control of economic systems, Chapter 5, optimization of traffic flow conditions, Chapter 6, or therapeutical actions for biological systems, Chapter 7.

4.5.1 Active Particles which Change Populations

Various systems of active particles are characterized by their ability to generate new particles in a population different from that of the interacting pairs. This is the case, as already mentioned, of particles which acquire a genetic structure different from that of the interacting pairs, Vogelstein and Kinzler (2004).

This subsection generalizes the mathematical structures developed in Section 4.4 to include the description of these phenomena, which are occasionally called **exotic interactions**, as seen in Chapter 3.

Shifting of populations may be related both to conservative and proliferative encounters. In the first case, the candidate particle changes population by falling into the population and microscopic state of the test particle; in the second case, proliferation of the candidate particle occurs, after an interaction with a field particle, in the state and population of the test particle. We do not discuss yet the conceivable applications; we simply offer mathematical structures to be used for modeling.

The analysis is developed only in the relatively simple case of spatially homogeneous systems, thus avoiding heavy notation. The generalizations to the space-dependent case involves additional calculations that can be developed with reference to specific applications.

The modeling of conservative interactions requires the following transition density:

$$\mathcal{B}_{rj}^{pq}(h, i), \qquad (4.5.1)$$

which denotes the discrete probability density that the candidate particle, with state p, of the rth population will fall, with state h, into the ith population after an interaction with the field particle, with state q, of the jth population.

The above probability density satisfies the normalization condition

$$\forall r, j, \, \forall p, q \, : \quad \sum_{h=1}^{H} \sum_{i=1}^{n} \mathcal{B}_{rj}^{pq}(h, i) = 1. \qquad (4.5.2)$$

The dynamics of exotic conservative interactions can be visualized using the representation of Figure 3.3.1 of Chapter 3.

Consider now nonconservative interactions. It is useful to split the term μ into the contribution of the proliferative and destructive parts, respectively, p and d. The term d is dealt with precisely as in Section 4.4 because a change of population can occur only for new particles, while destruction can only occur within the population of the test particle. The proliferation of a particle can be described by the following term:

$$p_{rj}^{pq}(h, i), \qquad (4.5.3)$$

which corresponds to proliferation, with state h, of a candidate particle with state p, of the rth population into the ith population of the test particle, after an interaction with the field particle, with state q, of the jth population. See Figure 3.3.2 of Chapter 3.

Calculations analogous to those we have seen above generate the following equation:

$$\frac{df_i^h}{dt} = \sum_{r=1}^{n}\sum_{j=1}^{n}\sum_{p=1}^{H}\sum_{q=1}^{H} c_{rj}\mathcal{B}_{rj}^{pq}(h,i)f_r^p f_j^q$$

$$+ \sum_{r=1}^{n}\sum_{j=1}^{n}\sum_{p=1}^{H}\sum_{q=1}^{H} c_{rj}\, p_{rj}^{pq}(h,i)f_r^p f_j^q$$

$$- f_i^h \sum_{j=1}^{n}\sum_{q=1}^{H} c_{ij}\bigl(1+d_{ij}\bigr)f_j^q, \qquad (4.5.4)$$

for $f_i^h = f_i^h(t)$.

The generalization to the spatially dependent case can be readily obtained using additional notation and calculations analogous to those developed in Section 4.4.

4.5.2 Active Particles with Totally Discrete Microscopic State

The previous analysis referred to systems with a microscopic state continuous in time, space, and velocity, but discrete in the activity. However, some systems of interest in the applied sciences need to be described by discrete variables also in space and/or velocity.

Let us consider the following discretization of the space and velocity microscopic variables:

$$I_{\mathbf{x}} = \{\mathbf{x}_1,\ldots,\mathbf{x}_\ell,\ldots,\mathbf{x}_L\}, \quad \text{and} \quad I_{\mathbf{v}} = \{\mathbf{v}_1,\ldots,\mathbf{v}_k,\ldots,\mathbf{v}_K\},$$

that have already been defined in Eqs. (4.3.14) and (4.3.18). Moreover, let us consider the distribution function, for the ith population, related to the point $\{\mathbf{x}_\ell, \mathbf{v}_k, \mathbf{u}_h\}$ of the space of microscopic states. This distribution is denoted by $f_i^{\ell k h}(t) = f_i(t, \mathbf{x} = \mathbf{x}_\ell, \mathbf{v} = \mathbf{v}_k, \mathbf{u} = \mathbf{u}_h)$, while the distribution function for the whole system is defined by the sum of Dirac distributions as in Eq. (4.3.19).

The mathematical framework is the evolution equation for the above discrete distribution function $f_i^{\ell k h}(t)$. The guidelines for its derivation are given below. We leave to the interested reader all technical calculations; however, detailed calculations will be developed in Chapter 5 for specific models of social dynamics.

The derivation of the mathematical framework can be obtained, as in the continuous case, according to the following guidelines:

1. Modeling the encounter rate so that it depends, for each pair of interacting populations, on the relative velocity of the interacting particles;

2. Modeling conservative interactions between the candidate and field particles depending on their microscopic states;

3. Modeling the proliferative terms related to nonconservative interactions between the test and field particles depending on their microscopic states;

4. Derivation of the evolution equation for each discrete distribution function using the conservation (balance) equation in the elementary volume of the space of the discrete microscopic states.

The above procedure is also valid for particles which have the ability of generating new particles in a population different from those of the interacting pairs. We immediately recognize that the structure of the evolution equation consists of a system of ordinary differential equations when the microscopic state is totally discrete, and it consists of a system of partial differential equations when the space variable is left continuous.

The strategy for selecting which of the microscopic variables needs to be discrete depends on the type of physical system to be modeled. The examples in the forthcoming chapters should clarify this.

4.5.3 Open Systems

Motivations to deal with open systems have already been given in Section 3.5 of chapter 3. The technical analysis essentially refers to the already cited paper by Brazzoli (2007).

We first propose the simple case of one population for systems where the microscopic state depends on the activity variable only. Subsequently, guidelines for generalizations to several interacting populations and to the spatially dependent case are given.

Let us consider the mathematical framework delivered by Eq. (4.4.8). The following specific actions are taken into account:

• $g(t) = \{g^q(t)\}$ is a stochastic action given, for each $q = 1, \ldots, H$, as a known function of time, acting over the candidate particle with state u^p. This action produces a shift into the state u^h of the test particle with probability density denoted by $\mathcal{C}^{pq}(h)$.

• $p(t) = \{p^q(t)\}$ is a stochastic action given, for each $q = 1, \ldots, H$, as a known function of time, acting over the test particle with state u^h. This action produces a proliferation or destruction in the state u^h of the test particle.

Calculations analogous to those we have seen in Section 4.4 generate the following framework:

$$\frac{df^h}{dt} = \sum_{p=1}^{H}\sum_{q=1}^{H} c^{pq}\mathcal{B}^{pq}(h)f^p f^q - f^h \sum_{q=1}^{H} c^{hq}(1 - \mu^{hq})f^q$$

$$+ \sum_{p=1}^{H}\sum_{q=1}^{H} c_e^{pq}\mathcal{C}^{pq}(h)f^p g^q - f^h \sum_{q=1}^{H} c_e^{hq} g^q$$

$$+ f^h \sum_{q=1}^{H} c_e^{hq} f^{hq} p^q \,, \tag{4.5.5}$$

where c_e^{pq} and c_e^{hq} denote the encounter rates of the candidate or test particles, respectively, with the external action.

Similarly to modeling with continuous variables, the case of deterministic actions can also be considered. Specifically, let $k^h(t)$ be a deterministic action, given as a known function of time, acting directly over the active particle with state u^h.

The following framework, which includes a linear proliferation or destruction effect, is obtained:

$$\frac{df^h}{dt} = k^h f^h + \sum_{p=1}^{H}\sum_{q=1}^{H} c^{pq}\mathcal{B}^{pq}(h)\,f^p f^q$$

$$- f^h \sum_{q=1}^{H} c^{hq} f^q + f^h \sum_{q=1}^{H} c^{hq} \mu^{hq} f^q \,. \tag{4.5.6}$$

The generalization to a system of several interacting population must consider the additive contribution of the interactions with all the different populations. The modeling of external actions causing shift or proliferation of particles in a population different from those of the interacting pair requires us to introduce additional terms related to the above actions.

Similar ideas can be developed for systems of particles whose microscopic state includes the space and velocity variables. Therefore, the external actions may also depend on these variables, and the generalization refers to the frameworks of Eqs. (4.4.9) and (4.4.10).

4.6 Critical Analysis

The contents of this chapter are motivated by the utility of modeling the microscopic state by a discrete variable rather than by a continuous one. Various examples of models in the forthcoming chapters motivate this approach.

It is not claimed, as in the preceding chapters, that the whole variety of conceivable formalizations has been considered. For instance, one possible a technical development is the generalization to the discrete case of the models with stochastic space dynamics in Section 3.4, to be used for the derivation of macroscopic equations. Several applications can be referred to these models.

The mathematical structures should be regarded as a formal framework. They can be specialized to model systems of interest in life sciences only if the various terms corresponding to microscopic interactions are properly modeled. The derivation of specific models may possibly suggest additional structures or generalizations.

As we have seen, the modeling of microscopic interactions is a crucial step which leads to the mathematical description of the evolution of the system. The structure of the mathematical frameworks derived in this chapter are technically different from those we have seen in Chapters 2 and 3. Specifically, the following classification is proposed.

I– Systems of particles with a microscopic state *identified only by the activity variable* are stated in terms of systems of *ordinary differential equations*, rather than the integro-differential equations needed by models with a continuous distribution over the microscopic state.

II– Spatially dependent models with a *continuous distribution over the velocity variable* generate systems of *integro-differential equations*, where the integral term refers to integration over the velocity variable, but not over the activity variable, while partial derivatives refer to the transport term in all space directions.

III– Spatially dependent models with *discrete microscopic velocity* generate systems of *partial differential equations*, rather than the systems of integro-differential equations of the continuous case. The integral term, Case II, is replaced by a finite sum, while partial derivatives refer to the transport term along the selected velocities.

IV– *Models with fully discrete microscopic state*, namely, space, velocity, and activity, are stated, as in Case I, in terms of systems of *ordinary differential equations*, because finite sums replace all integral terms.

These different mathematical structures require different computational methods to obtain simulations related to specific applications of models in the applied sciences.

The applications in the second part of the book refer to some of these schemes. For instance, the models analyzed in Chapter 5 correspond to Cases I and IV, while traffic flow models (Chapter 6) refer to Case III. The models analyzed in Chapters 7 and 8 are derived in the continuous framework, namely referring to Chapter 2. Deriving models with discrete microscopic structures to replace the continuous ones, and vice versa, is always an alternative to be considered in the modeling approach.

5

Modeling of Social Dynamics
and Economic Systems

5.1 Introduction

The mathematical tools developed in Chapters 2–4 are applied, starting
with this chapter, to the analysis of various complex systems of interest in
life and applied sciences—specifically to the modeling of the collective social
behavior of large systems of interacting individuals.

Interactions at the microscopic level have the ability to modify the social
state of the interacting pairs, while methods of generalized kinetic theory
can be used to describe the evolution of the probability distribution over
the microscopic state, which is identified by the social state. Moreover,
some research perspectives are proposed for modeling other complex social
systems: interactions and competition among nations, personal feelings,
and migration phenomena.

Referring to the formalism proposed in the previous chapters, the active
particles are now individuals, or groups of individuals, of a certain society.
The mathematical structures used to model the evolution of the system is
the one offered in Section 4.4, corresponding to a system with a constant
number of active particles characterized by a microscopic state identified
by the activity variable only.

This assumption means that the system is observed for short time intervals so that birth and death processes do not play a relevant role. Therefore,
the probability distribution over the microscopic state is, after a technical
normalization, a probability density. The modeling is developed in the case
of spatial homogeneity for systems where the microscopic state is discrete.
Moreover, various suggestions are given for developing suitable generaliza-

tions, such as modeling systems which include a space structure, or systems constituted by several interacting populations.

The idea of using methods of mathematical kinetic theory to study the social behavior of large populations of interacting individuals may be referred to the paper by Jager and Segel (1992), devoted to the modeling of the social behavior of certain populations of insects (the bumblebee). This model was introduced in Chapter 1. Microscopic interactions modify a particular social variable called **dominance**, which characterizes the ability of certain insects to dominate and organize the behavior of the other insects. The society, as experimentally observed by Hogeweg and Hesper (1983), evolves gradually towards a splitting into two particular classes: the dominant (a few individuals) and the dominated (the others).

Some ideas generalizing the contents of the above paper to the modeling of the social behavior of individuals in modern societies was introduced in Chapter 4 of the book by Bellomo and Lo Schiavo (2000), and a systematic analysis was developed in two recent papers by Bertotti and Delitala (2004), (2007), where the mathematical tools used in this chapter were originally proposed. These interesting papers offer the guidelines used to derive the models described in this chapter.

The approach to modeling by active particle methods is also due to Galam (2001), (2003) for models which describe the collective behavior of large systems of active particles based on their interactions with neighboring particles. An additional general bibliography is given in the last section of this chapter.

The models of this chapter refer not only to social systems but to the relatively more general framework of the *behavioral economy*, where deterministic rules of an economy may be stochastically perturbed by individuals' behaviors. Before approaching these mathematical topics, we analyze some conceptual aspects of the interaction between applied mathematics and social sciences. The main problem consists in understanding how the *qualitative interpretation* of social reality, delivered by research in the social sciences, can be transferred into a *quantitative description* produced by mathematical equations.

To achieve this, one needs the development of a dialogue between the methods and traditions of two different disciplines, with the additional difficulty of dealing with individuals who have the ability to think, and consequently to react to external actions without following deterministic causality rules. The dialogue, however difficult, is necessary.

This chapter is organized into four more sections as follows.

– Section 5.2 describes the scientific contributions by Bertotti and Delitala (2007), which is the reference model for this chapter. As we shall see, a critical analysis, with special attention to the interpretation of microscopic interactions, may generate relatively more general new models.

– Section 5.3 provides a qualitative analysis and some simulations related to the model described in Section 5.4.2. Particular attention is paid to the asymptotic behavior of the solutions, which is the main objective of the predictive ability of this class of models. The qualitative analysis generates a variety of interesting and challenging problems, which have already caught the interest of applied mathematicians.

– Section 5.4 discusses some perspective ideas and additional generalizations including the case of several interacting populations and microscopic states with a space structure. Particularly important are the economical implications of these generalizations, because migration events can be predicted within this approach. Various interesting problems are again brought to the attention of applied mathematicians.

These generalizations require some technical developments of the mathematical frameworks, e.g., referring to several interacting populations with transitions from one population to the other, but in a physical situation where the total number of active particles is constant in time.

– Section 5.5 investigates how the modeling approach developed in this chapter with reference to social dynamics can be properly generalized to other fields of social sciences, where microscopic interactions can modify the overall behavior of populations which are involved in a *game*, such as political orientations, and personal feelings and attitudes.

The physical interpretation of microscopic interactions proposed in Section 5.2 has a somewhat different meaning than those described in Chapters 2–4. Here the modeling refers to exchanges of social state, that is, the output of interactions somehow mediated by the economical organization of the society under investigation.

5.2 A Model by Bertotti and Delitala

This section describes a mathematical model proposed by Bertotti and Delitala (2007) devoted to modeling the behavior of the class of physical systems described in Section 5.1. As already mentioned, the modeling is developed using the mathematical structure developed in Section 4.4, which is reported again to make the contents of this section self-contained.

This model technically differs from a previous one proposed by the same authors (2004) in modeling the dynamics at the boundary of a microscopic state. Specifically, the model here refers to a closed system which does not allow, as we shall see, the lowest and highest classes to take part in the social competition; while the previous model allows all types of interactions, which is technically possible only in an open system.

The modeling refers to a large population of interacting individuals. Their microscopic state is described by the scalar variable $u \in [0, 1]$ which denotes the social state of the individuals, where $u = 0$ and $u = 1$ are, respectively, the lowest and the highest values of the social state. In particular, $u = 0$ refers to extreme poverty, and $u = 1$ to the highest social state. The modeling refers to a microscopic variable, with discrete values, defined as follows:

$$I_u = \{u_1 = 0, \ldots, u_h, \ldots, u_H = 1\}, \tag{5.2.1}$$

where H is the number of social classes or levels. Interactions modify the social state, but space phenomena are not taken into account. Namely, the space variable is not considered in assessing the microscopic state of the active particles.

The formal structure of the evolution equation used in the modeling process is as follows:

$$\frac{df^h}{dt} = J^h[\mathbf{f}] = \sum_{p=1}^{H} \sum_{q=1}^{H} \mathcal{B}^{pq}(h) f^p f^q - f^h \sum_{q=1}^{H} f^q, \tag{5.2.2}$$

for $h = 1, \ldots, H$, and where $\mathbf{f} = \{f^h\}$, $f^h = f^h(t)$, and all interaction rates are assumed to be equal to one and are hidden in the time scale. This technical simplification is, however, critically analyzed in the second part of the chapter.

The model is fully identified if the terms $\mathcal{B}^{pq}(h)$, which describe microscopic interactions, are properly modeled. These terms, called the **table of the game rules** in the original paper define the probability density (discrete) that a candidate individual with state p will fall into the state h of the test individual after an interaction with a field individual with state q. These terms have the structure of a discrete probability density:

$$\forall p, q = 1, \ldots, H : \quad \sum_{h=1}^{H} \mathcal{B}^{pq}(h) = 1. \tag{5.2.3}$$

The table of game rules is identified by H matrices with dimensions $H \times H$. In fact, an $H \times H$ matrix corresponds, for a fixed value h, to all conceivable encounters which generate the above output; this means that we need H different matrices.

Analogously, if p and q are fixed, then $\mathcal{B}^{pq}(h)$ identifies a string, corresponding to $h = 1, \ldots, H$, such that the sum of all elements is equal to one. Then, if p is fixed and q variable, the above mentioned strings identify an $H \times H$ matrix. Letting q take all values from 1 to H generates H matrices.

However, specific models reduce the number of parameters to a few as most of the entries of the matrices are common to all of them.

Considering that the number of individuals is constant in time, one has

$$\sum_{i=1}^{H} f^h(t) = \sum_{i=1}^{H} f^h(t=0) = f_0 . \qquad (5.2.4)$$

Therefore, all terms in (5.2.2) can be divided by f_0 so that all terms $f^h = f^h(t)$ assume the meaning of fraction of individuals, with respect to the total number of individuals, in the social state h. It follows that f can be regarded as a discrete probability density. The calculations developed in what follows take into account the above normalization which implies $f_0 = 1$.

As already mentioned, interactions have to be regarded not as effectively individual exchanges and/or competition, but as the output of complex mechanisms such as social conflicts, taxation politics, economical marketing, etc. Indeed, this remark supports the assumption that the variable u is discrete. Individuals related to the same value of u correspond to groups (or maybe classes) of individuals.

Before dealing with our specific mathematical models, we note that mathematical models can be divided into two main groups.

i) **Predictive models** have the ability to describe the evolution of a system starting from a given initial condition. These models involve parameters which have been well identified by comparisons with empirical data. In our case, models are derived after a detailed modeling, based on experimental information, of the table of game rules consisting of H matrices with dimensions $H \times H$.

ii) **Explorative models** are characterized by the aim of studying the trend of a system corresponding to different free parameters. A technical interpretation of the trend motivates to operate (or not), by a social politics, to promote certain types of social interactions. These models are based on tables of game rules that are proposed for explorative purposes. Their aim is to investigate the trend of the system corresponding to different types of interactions.

The specific model reported here can be interpreted as an explorative, rather than a predictive, model. It describes a social dynamics based on three parameters.

H: number of social classes;

$\gamma = d_c/H$: ratio of a suitable critical distance between social classes d_c with respect to their number;

L_0: total wealth of the whole population.

All three parameters play, as we shall see by analytic results and computations, a crucial role in the asymptotic behavior of the solutions. This interesting aspect is discussed in Section 5.3.

The dynamics of microscopic interactions is described by the following assumptions.

• Interactions within the same class $p = q$ do not modify the state of the interacting pairs:

$$\mathcal{B}^{pq}(h = q) = 1, \qquad \mathcal{B}^{pq}(h \neq p) = 0. \tag{5.2.5}$$

• Interactions between two individuals with sufficiently close social states, i.e., $|p - q| \leq d_c$, with $p \neq q$, generate the following competition: the individual placed in the higher social position improves its situation, while the one in a lower position faces a further decrease (***competitive behavior***),

$$p \quad \text{or} \quad q = 1, \quad \text{or} \, p \, \text{or} \, q = H \; : \; \begin{cases} \mathcal{B}^{pq}(h = p) = 1, \\ \mathcal{B}^{pq}(h \neq p) = 0, \end{cases} \tag{5.2.6a}$$

$$p < q, \quad p \neq 1, \quad q \neq H \; : \; \begin{cases} \mathcal{B}^{pq}(h = p - 1) = \alpha, \\ \mathcal{B}^{pq}(h = p) = 1 - \alpha, \\ \mathcal{B}^{pq}(h \neq p, p - 1) = 0, \end{cases} \tag{5.2.6b}$$

and

$$p > q, \quad p \neq H, \quad q \neq 1 \; : \; \begin{cases} \mathcal{B}^{pq}(h = p + 1) = \alpha, \\ \mathcal{B}^{pq}(h = p) = 1 - \alpha, \\ \mathcal{B}^{pq}(h \neq p, p + 1) = 0, \end{cases} \tag{5.2.6c}$$

where α is the fraction of individuals which are subject to a transition into a social state different from the departure one.

Therefore, in this type of competition, the lowest and highest social classes are frozen.

• Interactions between individuals with sufficiently distant social state, i.e., $|p - q| > d_c$, generate the opposite behavior (***altruistic behavior***). This assumption is formally written as follows:

$$p < q \; : \; \begin{cases} \mathcal{B}^{pq}(h = p + 1) = \alpha, \\ \mathcal{B}^{pq}(h = p) = 1 - \alpha, \\ \mathcal{B}^{pq}(h \neq p, p + 1) = 0, \end{cases} \tag{5.2.7a}$$

and

$$p > q \ : \ \begin{cases} \mathcal{B}^{pq}(h = p - 1) = \alpha \,, \\ \mathcal{B}^{pq}(h = p + 1) = 1 - \alpha \,, \\ \mathcal{B}^{pq}(h \neq p, p - 1) = 0 \,, \end{cases} \qquad (5.2.7b)$$

This dynamics is visualized in Figure 5.2.1.

h k h k

Fig. 5.2.1: Sketch of altruistic and competitive behavior.

Remark 5.2.1. *Substituting expressions (5.2.5)–(5.2.7) into the formal structure (5.2.2) generates specific models corresponding to the table of games. These models also preserve, in addition to the total number of individuals (5.2.4), the total wealth, which is defined as follows:*

$$\sum_{h=1}^{H} u^h f^h(t) = \sum_{h=1}^{H} u^h f^h(t = 0) = L_0 \,. \qquad (5.2.8)$$

Specific examples are given in what follows.

This class of models generates a variety of interesting mathematical problems concerning both analytic and modeling aspects. In detail, the qualitative analysis proposed in the original paper has shown that the solution to the initial value problem, obtained by linking Eq. (5.2.2) to initial conditions (5.2.4), uniquely exists for arbitrarily large times. A detailed study has been developed for the cases corresponding to $H = 5$ and $d_c = 2$; this result appears to have general validity considering that simulations developed for $H \geq 5$ confirm the behavior delivered by theorems.

The parameters which characterize the system have a well-defined physical meaning. Specifically, H is the number of social classes and d_c is the distance, actually referred to as H, between social classes which separates the competitive and the altruistic behavior of the individuals, and L_0 is the available wealth. The dynamics of a society also depends on this term:

this means that the asymptotic behavior of a rich country may substantially differ from that of a poor country, although the social organization to control the wealth is the same.

An immediate idea of the model is obtained by assuming special values of the parameters. Particularly interesting is the case of $H = 5$ and $d_c = 2$; using only five social classes already reproduces the qualitative behavior, produced by simulations, of models with a higher number of classes. Specifically, the model is

$$
\begin{cases}
\dfrac{df^1}{dt} = f^2 - f^{2^2} - f^1 f^2 - f^1 f^4 - f^1 f^5 - f^2 f^5 , \\[2mm]
\dfrac{df^2}{dt} = -f^2 + f^4 + f^{2^2} - f^{4^2} + f^1 f^2 - f^2 f^4 - f^4 f^5 + f^1 f^5 , \\[2mm]
\dfrac{df^4}{dt} = -f^4 + f^2 + f^{4^2} + f^{2^2} + f^5 f^4 - f^4 f^2 - f^2 f^1 + f^5 f^1 , \\[2mm]
\dfrac{df^5}{dt} = f^4 + f^{4^2} + f^5 f^4 + f^5 f^2 + f^5 f^1 + f^4 f^1 ,
\end{cases}
\tag{5.2.9}
$$

for $f^h = f^h(t)$, where f^3 has been eliminated by conservation of probability (5.2.4), and where the parameter α has been inserted into the time scale.

The global wealth can be expressed, using conservation of number of individuals, as a function ω of the four variables f_1, f_2, f_4, f_5:

$$
\omega : \Sigma \to \mathbb{R} , \qquad \omega(f_1, f_2, f_4, f_5) = -\frac{1}{2} f_1 - \frac{1}{4} f_2 + \frac{1}{4} f_4 + \frac{1}{2} f_5 , \tag{5.2.10}
$$

where

$$
\Sigma = \{(x_1, x_2, x_3, x_4) \in \mathbb{R}^4 : \quad x_j \geq 0
$$

$$
\text{for any} \quad j = 1, \dots, 4 \quad \text{and} \quad \sum_{j=1}^{4} x_j \leq 1 \} . \tag{5.2.11}
$$

It is immediate to show that the scalar function $\omega(f_1, f_2, f_4, f_5)$ is a first integral for the system (5.2.9). This property follows directly from the following relation:

$$
\frac{d\omega}{dt} = -\frac{1}{2} \frac{df_1}{dt} - \frac{1}{4} \frac{df_2}{dt} + \frac{1}{4} \frac{df_4}{dt} + \frac{1}{2} \frac{df_5}{dt} . \tag{5.2.12}
$$

Qualitative and computational analyses of the above explorative model can be be addressed to compute the asymptotic configurations (distributions of social classes) described by the model, so that the prediction of the model can be qualitatively analyzed. Despite its simplicity, the model already provides a tool to obtain useful investigations because its parameters can be related not only to social politics, but also to the behavior of the interacting individuals.

5.3 Qualitative Analysis and Simulations

The class of mathematical models described in Section 5.2 can be applied to the study of social systems after particularization of the interaction term $\mathcal{B}^{pq}$. Simulations can be obtained by solving the following initial value problem:

$$\begin{cases} \dfrac{df^h}{dt} = J^h[\mathbf{f}], \\[2mm] f^h(t=0) = f_0^h, \end{cases} \tag{5.3.1}$$

for $h = 1, \ldots, H$.

The model consists of a system of H ordinary differential equations with quadratic-type nonlinearities, which has to be solved with initial conditions f_0^h, where the set $\{f_0^h\}$ is a discrete probability density:

$$\sum_{h=1}^{H} f_0^h = 1. \tag{5.3.2}$$

The qualitative analysis of this problem has been developed by Bertotti and Delitala (2007), who have proved that, for any given set $\{f_0^h\}$ with $f_0^h \geq 0$ with $h = 1, \ldots, H$, the solution $f(t) = (f^1(t), \ldots, f^H(t))$ of problem (5.3.1) exists and is unique for all $t \in [0, +\infty)$.

In particular, it is shown that

$$\forall t \geq 0: \qquad f^h(t) \geq 0 \quad \text{for any } h = 1, \ldots, H, \tag{5.3.3}$$

and

$$\sum_{h=0}^{H} f^h(t) = 1, \quad \forall t \geq 0. \tag{5.3.4}$$

Moreover, there exists at least one equilibrium solution of the equation $J^h[\mathbf{f}]$ related to system (5.3.1).

The proof of the theorem of Bertotli and Delitala, which is based on classical fixed point arguments, is not reported here. This book is devoted to modeling and simulations rather than to proofs. However, it is simply remarked that the proof is obtained in two steps: the existence and uniqueness of the trajectories is first proved for any initial data which satisfy condition (5.3.4); while the second step, based on the Schauder fixed point theorem, proves the existence of at least one equilibrium point such that the right-hand side term of the first equation in (5.3.1) is equal to zero. Proof of positivity is simply obtained by exploiting the bilinear structure of the equation written in a suitable exponential form.

Uniqueness of equilibrium points and their stability properties has not been, until now, proved for general values of the parameters characterizing the model. A detailed analysis has been developed, in the above-cited paper, in the case

$$H = 5, \qquad d_c = \frac{H-1}{2}, \qquad \gamma = \frac{d_c}{H} = \frac{H-1}{2H}. \qquad (5.3.5)$$

It seems, after a systematic computational analysis, that this specific case can possibly act as a paradigm for several interesting more general cases.

The functional used for the proof is the following:

$$\mathcal{H} = \sum_{i=1}^{5} f_i \log f_i, \qquad (5.3.6)$$

which is proven to be not increasing along the solutions of (5.3.1). Moreover, its Lie derivative vanishes at the equilibrium points, while it is negative elsewhere.

The results of the qualitative analysis by Bertotti and Delitala (2007) are summarized in what follows to be used to define detailed objectives of simulations for models with different parameters.

Let us now consider the model with parameters defined in Eq. (5.3.5); the following results have been proven:

i) A unique equilibrium configuration exists for fixed values of the initial wealth L_0, which is preserved along the evolution;

ii) The internal equilibrium configuration (namely the configuration with each component different from zero) is asymptotically stable for any initial condition;

iii) The shape equilibrium configuration depends on L_0:

– For $L_0 = 0.5$, there is a trend towards a uniform distribution, over the various social classes, of the overall wealth;
– For $L_0 < 0.5$, there is a trend towards a distribution of the wealth concentrated over the low social classes; and for $L_0 > 0.5$, a trend towards a distribution of the wealth concentrated over the high social classes.

This result is visualized in Figures 5.3.1a,b,c which show the asymptotic equilibrium configuration for three different values of L_0.

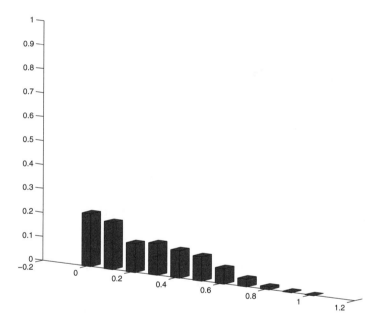

Fig. 5.3.1a: Wealth distribution at equilibrium: $L_0 = 0.25$, $d_c = 3$.

Figures 5.3.2a,b,c show the trajectories towards the asymptotic equilibrium configuration corresponding to the preceding three different values of L_0. Simulations show a monotone trend, which is typical of closed systems, while oscillations may be observed in the case of open systems depending on the behavior of the external actions. The trajectories corresponding to the various initial conditions are identified by a different graphic representation (dotted, dashed, etc.).

The general problem consists in showing how far the above results can be generalized to model the behavior of the system corresponding to different values of the parameters. Suitable simulations may provide useful indications and a hint towards the proof of additional theorems.

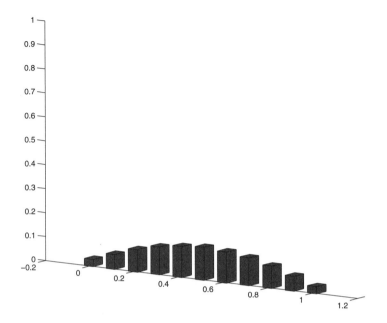

Fig. 5.3.1b: Wealth distribution at equilibrium: $L_0 = 0.5$, $d_c = 3$.

Simulations are developed in the case $H = 11$ with the aim of obtaining the following indications:

i) The influence of the parameters on the shape of the equilibrium configuration for $L_0 = 0.25$ and different values of d_c:

$$d_c < \frac{H-1}{2}, \qquad d_c = \frac{H-1}{2}, \qquad d_c > \frac{H-1}{2}; \qquad (5.3.7)$$

ii) The influence of the parameters on the shape of the equilibrium configuration in the cases defined in (5.3.7) for $L_0 = 0.5$ and $L_0 = 0.75$.

The results of the simulations are visualized in Figures 5.3.3a,b and 5.3.4a,b, corresponding, for each value of $L_0 = 0.25$, 0.75, to $d_c = 5$, and $d_c = 9$.

The trend to equilibrium is again monotone as previously shown in Figures 5.3.2a,b,c. Therefore, the simulations are not repeated.

The indications we can obtain from the above simulations are as follows:

- Large values of L_0, greater than the critical value $L_0 = 0.5$, correspond to a trend of the population to a distribution with the number of wealthy individuals larger than the number of poor individuals. The opposite

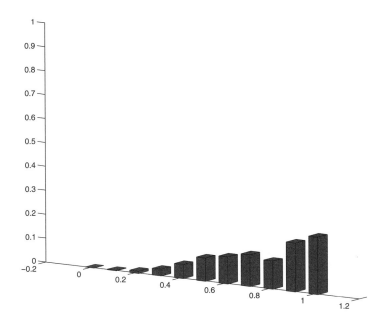

Fig. 5.3.1c: Wealth distribution at equilibrium: $L_0 = 0.75$, $d_c = 3$.

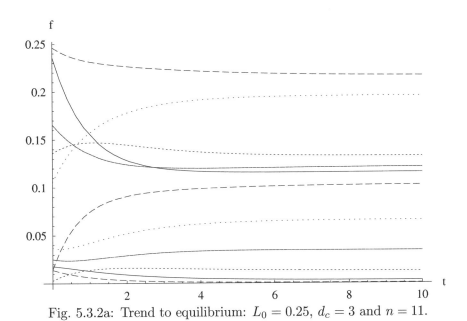

Fig. 5.3.2a: Trend to equilibrium: $L_0 = 0.25$, $d_c = 3$ and $n = 11$.

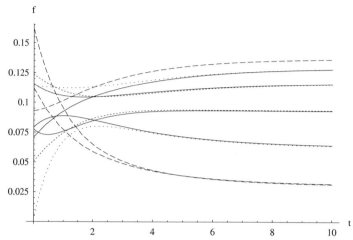

Fig. 5.3.2b: Trend to equilibrium: $L_0 = 0.5$, $d_c = 3$ and $n = 11$.

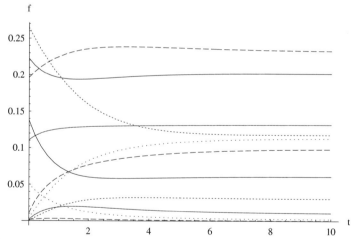

Fig. 5.3.2c: Trend to equilibrium: $L_0 = 0.75$, $d_c = 5$ and $n = 11$.

trend is described for small values of L_0, lower than the critical value $L_0 = 0.5$.

- The trend described for large values of L_0, greater than the critical value $L_0 = 0.5$, is contrasted by large values of the parameter d_c, while it is increased for small values of the parameter d_c.

- The trend described for small values of L_0, lower than the critical value $L_0 = 0.5$, is increased by large values of the parameter d_c, while it is contrasted for small values of the parameter d_c.

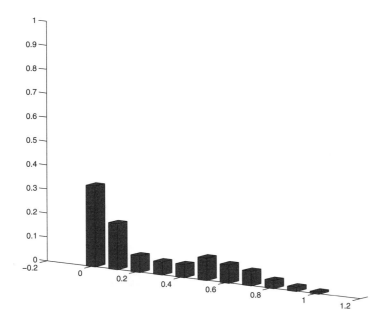

Fig. 5.3.3a: Equilibrium for $L_0 = 0.25$, $d_c = 5$.

Remark 5.3.1. *According to the model, the same social politics cannot be used, as often occurs, both for wealthy and poor countries. A proper choice of the parameter γ (or d_c if referred to a specific case) has to be selected to compensate the influence of L_0. Only rich countries can possibly be ruled by a competitive administration, while a welfare politics is necessary in poor countries to avoid a heavier concentration of lower classes.*

Remark 5.3.2. *The influence of L_0 refers to the total wealth which is effectively used in the economy. In other words, if part of the wealth is **hidden**, then the country shows a trend different from that one corresponding to the real wealth.*

Remark 5.3.3. *The model used in the simulations is based on the simplified assumption that the encounter rate does not depend on the microscopic state of the interacting pairs. However, the following, technically different, framework can be used:*

$$\frac{df^h}{dt} = J^h[\mathbf{f}] = \sum_{p=1}^{H} \sum_{q=1}^{H} c^{pq} \mathcal{B}^{pq}(h) f^p f^q - f^h \sum_{q=1}^{H} c^{hq} f^q, \qquad (5.3.8)$$

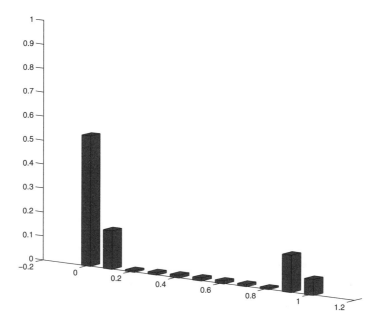

Fig. 5.3.3b: Equilibrium for $L_0 = 0.25$, $d_c = 9$.

where the interaction rates c^{pq} and c^{hq} depend on the microscopic states $\{u_p, u_q\}$ and $\{u_h, u_q\}$, respectively.

5.4 Some Ideas on Further Modeling Perspectives

The model of Sections 5.2 and 5.3 has shown how discrete kinetic equations for active particles can be used to obtain simulations of various phenomena related to social competition. This section provides some guidelines for modeling of additional phenomena and also for enlarging the approach to modeling systems technically different from the one in this chapter.

We remark that the mathematical framework described in Chapter 4 should be technically modified to enlarge the variety of phenomena related to the class of social systems dealt with in this chapter. Therefore, the mathematical tools offered in Chapter 4, summarized in the mathematical structures (5.2.2) and (5.3.8), can be regarded as a background, to be technically enlarged and developed, for modeling new particularized systems in social dynamics.

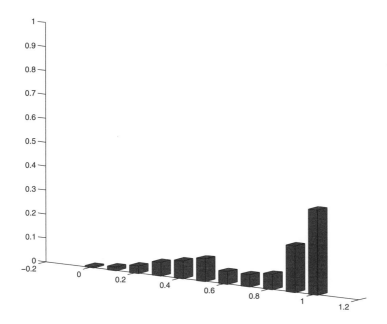

Fig. 5.3.4a: Equilibrium for $L_0 = 0.75$, $d_c = 5$.

The preceding sections have shown that the main issue is the study of the qualitative behavior of the solutions; specifically, the shape of the asymptotic behavior curve. Additional research is needed to generalize these results to the whole variety of parameters characterizing the model. Particularly interesting is the analysis of the influence of the overall wealth on the different shapes of the asymptotic behavior curves.

Apparently, theorems can be obtained only for specific cases corresponding to a low number of social classes, while computational analysis shows that the equilibrium distribution depends on the initial wealth which, according to the model, is preserved along the evolution. However, simulations show that the qualitative behavior is the same even if the number of social classes is increased.

Of course, simulations cannot be regarded as a mathematical proof, therefore a further development towards the proof of the above-mentioned qualitative analysis appears to be an interesting research perspective. Possibly, it can be applied to the models developed from the suggestions proposed in this section.

Referring to modeling aspects, this section proposes an introduction to some developments of the mathematical models described in Section 5.2 which are offered to the attention of applied mathematicians for research purposes. The various ideas that we will propose can possibly be developed

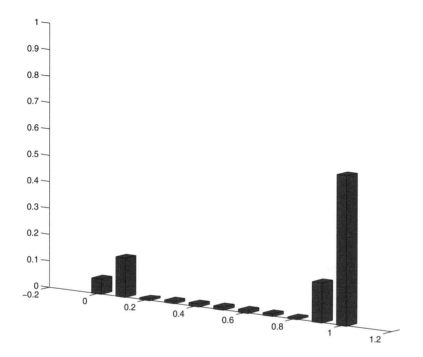

Fig. 5.3.4b: Equilibrium for $L_0 = 0.75$, $d_c = 9$.

towards the analysis of social systems more general than those of the previous sections. Specifically, the following topics, among others, are dealt with:

i) Social systems which involve more interacting populations, each of them characterized by different interaction rules within each population and pairs of individuals belonging to different populations;

ii) Microscopic states including systems with a vector microscopic activity, for instance, including the space variable to describe the localization of certain social classes and related migration processes;

iii) The modeling of systems subject to external actions which are organized to modify the evolution of the system and its trend towards the desired asymptotic configuration;

iv) Modeling systems where individuals shift from one population to the other due to modifications of their social state.

These topics are described in the four subsections which follow. Some concise indications to approach the above research perspectives are also given.

5.4.1 Models for Several Interacting Populations

Social systems may be constituted by more than one interacting population. Moreover, interactions within each population follow behavioral rules different from one population to the other, while analogous differences can be observed for different pairs of interacting populations. For instance, altruistic or competitive behaviors can be characterized by different parameters within each population, while the action of individuals of one population over the other may be totally competitive or totally altruistic.

Of course, different populations may be identified within the same society or even within the environment of different interacting nations. This type of investigation, which is missing in the literature, can possibly provide interesting descriptions certainly richer than those restricted to one population only. Moreover, different behaviors may be induced by social, political, and/or religious convictions.

The above situation corresponds to the behavior, sometimes present in our societies, where certain populations attempt to exploit other populations or to assist them in case of need.

The mathematical structure to deal with these systems refers again to Chapter 4. Specifically, we use the following class of equations:

$$\frac{df_i^h}{dt} = \sum_{j=1}^{n} \left(\sum_{p=1}^{H} \sum_{q=1}^{H} c_{ij}^{pq} \mathcal{B}_{ij}^{pq}(h) f_i^p f_j^q - f_i^h \sum_{q=1}^{H} c_{ij}^{hq} f_j^q \right), \qquad (5.4.1)$$

for populations $i = 1, \ldots, n$ and social classes $h = 1, \ldots, H$, and where $f_i^h = f_i^h(t)$ defines the discrete probability distribution function over the activity h in each ith population.

In this case, the interaction rates c_{ij}^{pq} and c_{ij}^{hq} are assumed to depend on the pair of interacting populations i, j and on the microscopic states of the interacting pairs p, q and h, q. Moreover, the **transition probability density** $\mathcal{B}_{ij}^{pq}(h)$ denotes the probability density that a **candidate** individual with state u_p, of the ith population will fall into the state u_h after an interaction with a **field** individual of the jth population with state u_q.

The above-defined transition density function has the structure, as in the case of one population only, of a probability density with respect to the variable u_h. In the discrete case with respect to the index h, this means

$$\forall i, j, \quad \forall p, q: \quad \sum_{h=1}^{H} \mathcal{B}_{ij}^{pq}(h) = 1. \qquad (5.4.2)$$

The encounter rate differs from one population to the next due to the fact that individuals within each population interact with a rate that differs in each population. Moreover, the interaction rate between individuals belonging to two different populations is generally of a lower order with respect to the one within the same population. In some cases, this rate can be equal to zero.

The qualitative analysis of the initial value problem, stated by linking to Eqs. (5.4.1) suitable initial conditions, provides from relation (5.4.2) the same global existence result we have seen in the case of one population only. The trend to equilibrium and stability properties can be studied by methods analogous to those used in Section 5.3. However, additional difficulties, which may be nontrivial, definitely arise.

Particularly interesting is the specific case, not yet studied in the literature, of two interacting populations. The model can possibly show an asymptotic trend which differs from one population to another. For instance, the wealth within the first population may be obtained by a progressive exploiting of the second one. On the other hand, a correct political and/or social organization of microscopic interactions can generate a trend such that the two populations, despite some differences characterizing their behaviors, may reach equilibria corresponding to comparably equal social states.

An additional phenomenon to be studied is the influence of the wealth of each population on the shape of the asymptotic solution. As we have seen, different shapes correspond to different values of the wealth. Therefore, even if the wealth within each population is preserved, one population may evolve towards states with a fair distribution of the wealth, while another population shows a trend towards an unfair and undesired wealth distribution.

A further interesting case refers to models where the overall wealth is preserved, while a transfer of wealth occurs from one population to the other. The following questions can be posed.

i) When two different populations (nations) interact, is it possible that one of the two shows a trend towards a fair distribution of the social state, while the other shows an unfair trend?

ii) When the trend described in item (i) occurs, does the wealth shift from one population to the other, while the overall wealth is preserved?

iii) Does the model describe an oscillation, in time, of the trend described in item (i)?

iv) How must the parameters modeling microscopic interactions be modified to control the trend described in item (i)?

v) How can external actions modify or even control such a trend?

Possibly, the answer to these questions should be delivered not only by simulations, but also by a rigorous qualitative analysis of the initial value problem.

5.4.2 Models with Vector Microscopic Activities

The class of models discussed until now have been characterized by the fact that the microscopic state of the interacting individuals is a scalar variable. However, modeling may require enlarging the dimension of the microscopic variable with the aim of providing a relatively more precise description of the state of the interacting individuals. For instance, the social state is not simply related to economical wealth, but also to other variables, e.g., education or political opinions. An additional component of the microscopic variable is the space localization corresponding to social states, because modifications of the social state (or certain social states) may induce migration phenomena.

Also, the previous mathematical models have been derived in the spatially homogeneous case, which is consistent with a certain interpretation of physical reality, considering that the velocity of individuals is not a meaningful variable as in mechanics. But it can be of some interest to introduce in the microscopic state an additional variable corresponding to the localization of individuals, who may move from one area to another due to migrations related to their social state. Namely, modifications of the social state may induce migrations from one area to the other, while the number of individuals in a certain area motivates emigration or immigration.

Let us concentrate on the case of a microscopic state which includes the localization of the individuals. Moreover, let us consider the case of one population only. Enlarging the microscopic state to introduce the above-defined localization means dealing with the probability density function:

$$f^{hk}(t) = f(t, u_h, x_k),\qquad (5.4.3)$$

where x_k, with $k = 1, \dots, K$ identifies the localization.

It is understood that localization is regarded as an independent microscopic variable, while velocity is not considered to be descriptive of the state of interacting active particles. This is due to the fact that localization does not have a physical meaning but identifies socially characterized areas.

The mathematical structure suitable to describe the evolution of the above probability density is

$$\frac{df^{hk}}{dt} = \sum_{p=1}^{H}\sum_{q=1}^{K}\sum_{s=1}^{H}\sum_{r=1}^{K} c_{sr}^{pq} \mathcal{B}_{sr}^{pq}(hk) f^{sr} f^{pq} - f^{hk} \sum_{q=1}^{H}\sum_{s=1}^{K} c_{qs}^{hk} f^{qs},\qquad (5.4.4)$$

where:

   hk represents the state of the test individual;

   pq represents the state of the field individual;

   sr represents the state of the candidate individual;

$\mathcal{B}_{sr}^{pq}(hk)$ defines the probability density that a candidate individual with state sr will fall into the state hk after an interaction with an individual with state pq:

$$\forall p,q, \quad \forall s,r: \quad \sum_{h=1}^{H}\sum_{k=1}^{K}\mathcal{B}_{sr}^{pq}(hk) = 1. \qquad (5.4.5)$$

An alternative to this approach consists in increasing the number of populations from one to K, namely by referring each population to a scalar activity only. The number of ordinary differential equations turns out to be $H \times K$.

The above type of modeling should be technically developed for systems with more than one interacting population. Migration phenomena are very interesting to analyze in the case of a population keen to maintain stable localization, and others keen to migrate. If the activity is reduced to a scalar variable, the number of equations is $n \times H \times K$.

The same reasoning developed at the end of the preceding subsection, concerning the analysis of several interacting populations, can also be applied in this case which includes the possibility of studying migration phenomena.

5.4.3 Modeling Open Systems

The preceding mathematical models do not include the description of external actions which can be applied to modify the evolution of the system. Their modeling can be developed using a mathematical structure analogous to the one of the autonomous system. The reasoning to be followed is that of Chapter 4, which again is adapted here.

Let us consider the following set of variables:

$$I_v = \{v_1 = 0,\ldots,v_j,\ldots,v_H = 1\}, \qquad (5.4.6)$$

which has the same dimension as the variable characterizing the microscopic state of the active particles.

Suppose now that the variables defined in Eq. (5.4.6) are linked to a known discrete probability density:

$$\mathbf{g} = \{g^j(t)\}, \qquad (5.4.7)$$

such that

$$\forall t \geq 0: \quad \sum_{j=1}^{H} g^j(t) = 1. \qquad (5.4.8)$$

Therefore, supposing that the action **g** acts on **f** with rules analogous to those on the interactions between active particles, the formal structure of the evolution equation, for a system constituted by one population only, is as follows:

$$\frac{df^h}{dt} = \sum_{p=1}^{H}\sum_{q=1}^{H} c^{pq}\mathcal{B}^{pq}(h)f^p f^q - f^h \sum_{q=1}^{H} c^{hq}f^q$$

$$+ \varepsilon \sum_{p=1}^{H}\sum_{q=1}^{H} c^{pq}\mathcal{G}^{pq}(h)f^p g^q - \varepsilon f^h \sum_{q=1}^{H} c^{hq}g^q, \qquad (5.4.9)$$

for $h = 1, \ldots, H$, and where ε denotes the time scaling of the interaction rates between particles and external actions with respect to the interaction rate between the particles themselves. This parameter may also include the modeling of the intensity of the action.

As we can see, the modeling of the external action over the particles is identified if the terms $\mathcal{G}^{pq}(h)$, which may be called the **table of the game actions**, define the probability density (discrete) that a candidate individual with state p will fall into the state h of the test individual after an interaction with an external agent with state q.

The terms $\mathcal{G}^{pq}(h)$, analogously to $\mathcal{B}^{pq}(h)$, have the structure of a probability density:

$$\forall p, q = 1, \ldots, m : \quad \sum_{h=1}^{H} \mathcal{G}^{pq}(h) = 1. \qquad (5.4.10)$$

In general, this model shows some technical analogy to a system of two populations. The first population corresponds to the active particles, and the second one to the external actions which are constant in time, or given as known functions of time. This explains why the interaction rates differ. The generalization to the case of several interacting populations requires additional technical calculations to include all contributions related to interactions between active particles of different populations.

5.4.4 Models with Shift of Populations

In some cases, for instance, in systems consisting of several interacting populations, interactions at the individual (microscopic) level may induce a shift from one population to the other. This shift may be related to an increase or decrease of wealth, for instance, racial discrimination appears to be less relevant for wealthy people.

A simple immediate example refers to the analysis of systems constituted by two interacting populations: the first one consists of local inhabitants, while the second one consists of immigrants. A rough classification (which may be valid at least initially) is related only to the social state. Then, interactions may cause transitions from one population to the other related to a certain threshold of the social state. For instance, individuals of the first population whose state is below the threshold fall into the second population, while individuals of the second population whose state goes over this threshold fall into the first population.

Only a rough indication has been given to provide a preliminary idea of the modeling perspectives. These would be developed only after a deeper analysis of the complexity of the system. In this case, the mathematical structure is the same as the one in Chapter 4.

5.5 Critical Analysis and Further Developments

The class of physical systems analyzed in this chapter can be regarded as an interesting field of investigation due to its potential impact on social sciences. Predicting the evolution of social systems starting from individual interactions is a fascinating idea which has always attracted applied mathematicians and physicists, as documented in various scientific papers, e.g., Galam (2003), (2004), Lo Schiavo (2003), (2006), and many others. The general background is that of the science of economics viewed as a complex evolutionary system, as documented in the lectures edited by Arthur, Durlauf, and Lane (1997).

The approach to understanding economical systems as a network of interacting agents, where individual behaviors play a significant role, has involved several researchers in the field of economics, among others, Krugman (1996), Arthur (1999), Asselmeyer, Ebeling, and Rosé (1997), Axelrod (1997), Batty and Xie (1994), Frankhauser (1994), Helbing (1995), Portugali (2000), Schweitzer (1998), and Weidlich (2000).

The class of mathematical models described in Section 5.2 is characterized by interaction rules, at the microscopic level, depending only on the microscopic state of the interacting pairs. Moreover, interactions are such that all individuals in a certain state change their state due to the interactions. Additional structures have been developed in Section 4.4.

The modeling has been developed within the frameworks proposed in Chapter 4 for systems with a discrete microscopic state. The motivation to use this type of mathematical structure, rather than the one for a continuous microscopic state, has been given in Section 5.1. The analysis of this chapter can be, however, straightforwardly generalized to the continuous case. A technical advantage of the continuous modeling is that the identification of the lower and higher social classes is not constrained.

However, the modeling above can be improved considering that the various phenomena, which have been constrained into a mathematical framework, are characterized by several complexity features. For instance, the output of microscopic interactions may depend on the number of individuals in the social classes involved in the social exchanges. In this case, the simple bilinear form used in this chapter may be replaced by more complex frameworks.

Moreover, in some cases, multiple interactions have to be properly modeled. This is not a simple task, as documented in the paper by Bellomo and Carbonaro (2006) devoted to modeling psychological feelings in a game involving three persons. The need of including triple interactions in the table of games was also posed by Platkowski (2004). The complexity of modeling increases consistently, as analyzed in Chapter and as shown in this chapter in attempting to develop a mathematical approach, or even some mathematical tools, to deal with the modeling of social systems.

Of course, including the above descriptive ability is not simply a matter of technical development of the mathematical structures offered in Chapters 2–4. It requires a deep insight into the above-mentioned structure towards the derivation of new ones certainly characterized by higher degree of complexity. Future research activity will definitely improve these tools and possibly derive new mathematical structures.

All these ideas need additional refining when applied to systems of several interacting populations in the case of open systems. In this case, additional analysis is needed, and the descriptive ability of the models needs to tackle relatively more complex, but also more interesting, systems.

The challenging difficulties also include the qualitative analysis of the initial value problem where a variety of analytic problems are still open. Particularly interesting is the study of models for open systems finalized to identify the asymptotic, in time, behavior of the solutions. This can possibly be referred to optimization and control problems.

Finally, let us discuss again the strategy of selecting a model with discrete states with respect to modeling the microscopic state by a continuous variable. Such a framework has been chosen according to the idea that social classes can be properly identified by a discrete variable corresponding to ranges of the social state, rather than by a continuous variation.

However, a technical drawback appears: the model refers to a fixed number of social classes, while in social systems a new social class may appear. The mathematical structure can be technically developed to include the above phenomena. However, the model appears to be a system of equations where their number evolves in time. However, a continuous distribution can be used to overcome this technical difficulty.

Let us consider then a social system where the social state is identified by a variable $u \in \mathbb{R}$, where the value $u = 0$ separates the acceptable (positive) social states from the poverty identified by negative values. The overall state of the system is described by the distribution function

$$f = f(t, u) : \quad [0, T] \times \mathbb{R} \to \mathbb{R}^+ \quad f \in L_1(\mathbb{R}), \tag{5.5.1}$$

where

$$\lim_{|u| \to \infty} f = 0, \qquad \int_{\mathbb{R}} f(t, u) \, du = 1, \quad \forall \, t \geq 0. \tag{5.5.2}$$

The mathematical structure to be used for modeling closed systems is as follows:

$$\frac{\partial}{\partial t} f(t, u) = \int_{\mathbb{R} \times \mathbb{R}} \eta(u_*, u^*) \mathcal{B}(u_*, u^*; u) f(t, u_*) f(t, u^*) \, du_* \, du^*$$

$$- f(t, u) \int_{\mathbb{R}} \eta(u, u^*) f(t, u^*) \, du^*. \tag{5.5.3}$$

Moreover, the structure for open systems is written, with an immediate meaning of the notation, as follows:

$$\frac{\partial}{\partial t} f(t, u) = \int_{\mathbb{R} \times \mathbb{R}} \eta(u_*, u^*) \mathcal{B}(u_*, u^*; u) f(t, u_*) f(t, u^*) \, du_* \, du^*$$

$$- f(t, u) \sum_{j=1}^{n} \int_{\mathbb{R}} \eta(u, u^*) f(t, u^*) \, du^*$$

$$+ \varepsilon \int_{\mathbb{R} \times \mathbb{R}} \eta(u_*, u^*) \mathcal{G}(u_*, u^*; u) f(t, u_*) g(t, u^*) \, du_* \, du^*$$

$$- \varepsilon f(t, u) \int_{\mathbb{R}} \eta(u, u^*) g(t, u^*) \, du^*. \tag{5.5.4}$$

If the modeling refers to several interacting populations, the following structures can be used:

$$\frac{\partial}{\partial t} f_i(t, u) = \sum_{j=1}^{n} \int_{\mathbb{R} \times \mathbb{R}} \eta_{ij}(u_*, u^*) \mathcal{B}_{ij}(u_*, u^*; u) f_i(t, u_*) f_j(t, u^*) \, du_* \, du^*$$

$$- f_i(t, u) \sum_{j=1}^{n} \int_{\mathbb{R}} \eta_{ij}(u, u^*) f_j(t, u^*) \, du^*, \tag{5.5.5}$$

and

$$\frac{\partial}{\partial t} f_i(t, u) = \sum_{j=1}^{n} \int_{\mathbb{R} \times \mathbb{R}} \eta_{ij}(u_*, u^*) \mathcal{B}_{ij}(u_*, u^*; u) f_i(t, u_*) f_j(t, u^*) \, du_* \, du^*$$

$$- f_i(t, u) \sum_{j=1}^{n} \int_{\mathbb{R}} \eta_{ij}(u, u^*) f_j(t, u^*) \, du^*$$

$$+ \varepsilon \sum_{j=1}^{n} \int_{\mathbb{R} \times \mathbb{R}} \eta_{ij}(u_*, u^*) \mathcal{G}_{ij}(u_*, u^*; u) f_i(t, u_*) g_j(t, u^*) \, du_* \, du^*$$

$$- \varepsilon f_i(t, u) \sum_{j=1}^{n} \int_{\mathbb{R}} \eta_{ij}(u, u^*) f_j(t, u^*) \, du^* . \tag{5.5.6}$$

The modeling of microscopic interactions can be developed following the same guidelines of the approach described in Section 5.2. The simulations of the initial value problem may eventually provide technically different results, while the qualitative analysis should tackle all difficulties involved by the new structure, which includes operators with different properties referring to the infinite domain of the microscopic variable.

In conclusion, the above mathematical structures can offer a useful background to describe more general economic systems where individual behaviors and interaction play a role over the evolution of the system. In some cases, these behaviors and interactions can play a role over the output of conflicts and competition.

The works reported in Section 5.1 indicate how economists place a great relevance on the role that individual behaviors and interactions can play in the evolution of economical systems. Of course, different approaches can possibly be used that may be related to scaling for economical systems. Several models corresponding to a macroscopic description, delivered by dynamical systems theory, are reported in the book by Wei-Bin Zhang (2005).

In general, random behavior must be taken into account. Therefore, the book by Capasso and Bakstein (2005) offers a valuable survey of methodological tools in the field of stochastic processes with applications to the study of economical systems.

This final section has proposed various ideas for research perspectives. Our ideas are simply a hint towards future research speculations. The most relevant objective remains the development of a proper mathematical theory for economical systems.

6

Mathematical Modeling
of Vehicular Traffic Flow Phenomena

6.1 Introduction

The modeling of systems treated in Chapter 5 refers to the case of spatial homogeneity (being well mixed), and also to a number of interacting active particles that are constant, or known, in time. This chapter deals with systems with a somewhat different structure: the number of active particles is still constant in time, but the particles are heterogeneously distributed in space. Specifically, the modeling refers to the mathematical description of traffic flow phenomena for vehicles in one- or multi-lane roads. If the inlet and outlet of vehicles are given, the number of vehicles included in the road can be regarded as a known function of time.

Vehicles should be modeled as active particles because their mechanical properties need to be integrated by considering the role of the driver, which differs from vehicle to vehicle. Drivers may be experienced or inexperienced, timid or aggressive, and so on. Even the properties of vehicles differ from case to case, and consequently their specific characteristics cannot be considered constant parameters of the system. Possibly, one may deal with them by using random variables.

The above reasoning suggests that we include, in the description of the microscopic state, also an internal *activity* variable suitable to model the specific features of the driver-vehicle system, from slow vehicles *linked* to an inexperienced driver to a fast vehicle with an experienced driver.

The greatest part of the existing literature (see Klar, Kühne, and Wegener (1996)), is based on models which describe interaction between vehicles without taking into account the role of the activity variable. This chapter reports on the existing scientific contributions, but it also introduces the modeling of traffic phenomena by methods of the kinetic theory of active particles.

The literature in the field, despite the novelty of this specific topic, is vast due to various social and ecological motivations. However, it does not yet provide a satisfactory answer to some crucial topics. Specifically:
i) The selection of the appropriate representation scale (microscopic, macroscopic, or statistical);
ii) The assessment of the variables which identify the activity, and hence the interactions between the activity and mechanical dynamics.

This chapter attempts to introduce the reader to a mathematical approach suitable for dealing with the above-mentioned modeling aspects. The analysis, although essentially devoted to methods of mathematical kinetic theory, also refers to microscopic and macroscopic modeling, so that an overall overview on traffic flow modeling is given.

The chapter is organized through five more sections.

– Section 6.2 introduces the concepts of scaling and representation of traffic flow phenomena. Different classes of equations correspond to each type of representation. A concise introduction is given to the mathematical frameworks corresponding to the microscopic and macroscopic scales which are also used as a reference for the kinetic modeling.

– Section 6.3 provides a concise description of a selection of various models available in the literature according to the framework of classical mathematical kinetic theory, thus showing how the modeling approach has evolved in time. The essential bibliography in the field is reported, showing how the the approach has been developed, however without yet taking into account the activity variable.

The survey does not claim to be exhaustive, but simply to provide a general overlook by means of selected examples. This section also analyzes various mathematical frameworks to be used for different modeling approaches.

– Section 6.4 discusses modeling by discrete velocity models. A discrete velocity variable is used to overcome the criticism on the assumption of continuity over the microscopic state that is required by classical kinetic theory. The discussion is related to a model proposed by Coscia, Delitala, and Frasca (2007), where the discretization of the velocity variable is related to the local density conditions. This section also provides some perspective ideas on discretization of the whole phase space.

– Section 6.5 reports on an alternative to the above model as documented in the paper by Delitala and Tosin (2007), where it is shown how simulations are able to reproduce physical phenomena observed by empirical data. The alternative essentially rests on the fact that the test and candidate particles interact with the field particles in a domain corresponding to the visibility zone of the driver. A relevant feature of this model is that a detailed mathematical description of microscopic interactions generates a model able to reproduce macroscopic behaviors resulting from experimental data. A few simulations visualize some interesting descriptions of real flow phenomena in agreement with observed data.

– Section 6.6 identifies the activity variable related to the modeling of the interactions between the driver and the dynamics of the vehicle. Subsequently, a methodological approach for the application of methods of mathematical kinetic theory for active particles is developed. The reasoning refers specifically to the class of models described in Sections 6.4 and 6.5. Moreover, this section analyzes some research perspectives starting from a general critical analysis to the traditional approaches to traffic flow modeling.

The book by Kerner (2004) is a constant reference for this chapter. This book provides a detailed and sharp interpretation of the physics of traffic phenomena, thus focusing various observed events which should be properly described by mathematical models. The book reports a variety of specific phenomena which are observed in traffic flow conditions and provides a general framework, from the viewpoint of physics, that defines a valuable background to be used for modeling.

The interpretation of the physics of traffic flow and the qualitative description of specific phenomena is an essential reference for mathematical modeling. Indeed, models must possess the ability to describe the qualitative behavior of specific traffic flow phenomena, because comparisons at a quantitative level may be very difficult, or even impossible, to achieve due to the randomness and irregularity of traffic flow.

6.2 Scaling and Representation

Traffic flow phenomena, like almost all systems of the real world, can be observed and represented at different scales. Specifically, the following types of descriptions have been used.

Microscopic description: refers to vehicles individually identified. In this case, the position and velocity of each vehicle define the state of the system as dependent variables of time. Mathematical models describe their evolution generally by systems of ordinary differential equations.

Kinetic theory description: used when the state of the system is still identified by the position and velocity of the vehicles, however, their identification does not refer to each vehicle, but to a suitable probability distribution over the microscopic state regarded as a random variable. Mathematical models describe the evolution of the distribution function using nonlinear integro-differential equations of the type we have seen in the preceding chapters.

Macroscopic description: used when the state of the system is described by averaged gross quantities, namely density, linear momentum, and kinetic energy of the vehicles, regarded as dependent variables of time and space. Mathematical models describe the evolution of these variables by systems of partial differential equations. Generally, models are limited to the first two quantities, because the energy may be quite difficult to model by a macroscopic description.

Hybrid models: can be developed by cellular automata techniques as a device to perform local averages.

Although this chapter is devoted to modeling by kinetic theory methods, some information on modeling at the microscopic and macroscopic scales can be useful both for the interpretation of experimental results and for modeling at the other scales. Microscopic models can contribute to the mathematical description of the interactions between vehicles in the models obtained by kinetic theory, while macroscopic models must be related to the equations delivered by suitable asymptotic methods applied to the kinetic equations.

However, modeling the above complex systems has the additional difficulty that none of the usual representation scales is effectively consistent with the physics of the system itself. This matter is properly discussed in Subsection 6.2.4.

Some preliminary definitions are useful for the analysis developed in what follows. Let us then consider a one-directional flow of vehicles along a road with length ℓ with one or more lanes each labeled by the superscript j, where $j = 1, \ldots, R$. For all lanes time and space variables are the independent variables. Specifically:

• t is the dimensionless time variable obtained by referring the real time t_r to a suitable critical time T_c to be properly defined by a qualitative analysis of the differential model;

• x is the dimensionless space variable obtained by referring the real space x_r to the length ℓ of the lane.

Moreover, it is useful to define the dependent variables, for each of the above representation scales, introducing suitable reference values:

n_M is the maximum density of vehicles corresponding to a bumper-to-bumper traffic jam;

V_M is the maximum admissible mean velocity which can be reached, on average, by vehicles running in free flow conditions.

Of course, a fast isolated vehicle can reach velocities larger than V_M. In particular, a limit velocity can be defined as follows:

$$V_\ell = (1 + \mu)V_M\,, \qquad \mu > 0\,, \tag{6.2.1}$$

such that no vehicle can reach, simply for mechanical reasons, a velocity larger than V_ℓ.

It is convenient, as we shall discuss with reference to macroscopic modeling, to identify the critical time T_c as the ratio between ℓ and V_M.

The selection of the dependent variable suitable for the representation of the system can be organized for each particular scale as shown in the three subsections that follow. A specific class of equations corresponds to each scale. The analysis is developed without introducing the activity in the microscopic state, however, the active particle mathematical approach is analyzed in Section 6.6.

6.2.1 Microscopic Models

When all vehicles are individually identified, the state of the whole system is defined, for each lane, by the dimensionless position and velocity of the vehicles which can then be regarded as points:

$$x_i = x_i(t)\,, \qquad v_i = v_i(t)\,, \qquad i = 1, \ldots, N\,, \tag{6.2.2}$$

where the subscript refers to each vehicle, and where $x_i \in [0, 1]$ and $v_i \in [0, 1+\mu]$ are dimensionless variables being referred to ℓ and V_M respectively.

This representation refers to a one-lane flow, where a superscript is necessary to identify the lane. In this case, the variables x_i^j and v_i^j are defined for $i = 1, \ldots, N$ and $j = 1, \ldots, R$.

The knowledge of these quantities can provide, using suitable averaging processes, gross quantities such as density and mass velocity. However, this is a delicate problem related to the fact that the real discrete system, made up of single vehicles, has been approximated by a continuous flow. Therefore, an averaging needs to be performed; see Darbha and Rajagopal (2002), and Tyagi, Darbha, and Rajagopal (2007). In principle, these physical quantities can be averaged either at fixed time over a certain space domain or at fixed space over a certain time range. For instance, the number density is given, for each lane, by the number of vehicles $n(t)$ which at time t are found in the domain $[x - \Delta, x + \Delta]$, as follows:

$$\rho(t; x) \cong \frac{1}{2\Delta} \frac{n(t; x)}{n_M}\,. \tag{6.2.3}$$

Calculations can also be related to each lane; then the density in each lane is denoted as $\rho^j(t; x)$, see Figure 6.2.1.

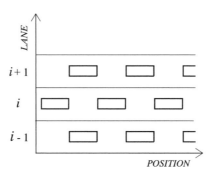

Fig. 6.2.1: Multilane flow.

Similar methols can be applied for the mass (mean) velocity

$$\xi(t;x) \cong \frac{1}{\rho(t;x)} \sum_{i=1}^{n(t;x)} v_i(t),\qquad(6.2.4)$$

where $v_i(t)$ denotes the velocity of the ith vehicle, at time t in the domain $[x - \Delta, x + \Delta]$, and $n(t;x)$ is the number of vehicles in the tract. However, the choice of the space interval is a critical problem and fluctuations may be generated by different choices of Δ.

If the above representation is particularized for each j-lane, summing over all lanes gives the local quantities in the whole road:

$$\rho(t;x) = \sum_{j=1}^{R} \rho^j(t;x),\qquad(6.2.5)$$

and

$$V(t;x) = \sum_{j=1}^{R} v^j(t;x)\cdot\qquad(6.2.6)$$

The averaging can be developed in a time interval rather than in a space interval. This is, in some cases, practically related to experimental measurements. However, fluctuations cannot be avoided.

Mathematical models at the microscopic scale have a structure analogous to that of Newtonian dynamics. The model describes the acceleration of vehicles as the output of the action of surrounding vehicles. In general, for a one-lane flow, the structure of models is as follows:

$$\begin{cases} \dfrac{dx_i}{dt} = v_i \,, \\[2mm] \dfrac{dv_i}{dt} = F_i(x_i, \ldots, x_N, v_i, \ldots, v_N) \,, \end{cases} \tag{6.2.7}$$

where $i = 1, \ldots, N$, and F_i is the acceleration acting on the ith vehicle. In general, F_i depends on the position and velocity of all vehicles.

Due to the complexity in the mathematical description of F_i, specific models simply relate the acceleration of the vehicle to the action of the **leader**, namely the vehicle in front of the **test** vehicle. The above assumption already shows the need to simplify the technical difficulties of the modeling process. As a matter of fact, a driver organizes the dynamics of the vehicle in its visibility zone. However, the difficulty in dealing with a large system of ordinary differential equations obliges one to reduce the complexity of the dynamics of each vehicle.

6.2.2 Macroscopic Models

As we have seen, macroscopic quantities can be recovered by local averages of the microscopic state of active particles, in our case vehicles. The representation by macroscopic models uses the above quantities directly. For instance:

$\rho = \rho(t, x)$ is the dimensionless density referred to the maximum density n_M of vehicles;

$\xi = \xi(t, x)$ is the dimensionless mean velocity referred to V_M.

The mean velocity can be replaced, if useful, by the flow:

$q = q(t, x)$, that is the linear momentum referred to $q_M = n_M V_M$.

The mathematical framework for the design of specific models at the macroscopic scale is based on mass and momentum conservation equations obtained assuming that the flow of vehicles is continuous with respect to both dependent variables. Hence, one has two independent coupled equations: the first one for mass conservation, the second one for momentum.

Using dimensionless variables, after selecting the reference time as follows:

$$\frac{T_c V_M}{\ell} = 1 \quad \Rightarrow \quad T_c = \frac{\ell}{V_M} \,, \tag{6.2.8}$$

leads, in the absence of source terms, to the following structure:

$$\begin{cases} \dfrac{\partial \rho}{\partial t} + \dfrac{\partial}{\partial x}(\rho \xi) = 0 \,, \\[3mm] \dfrac{\partial \xi}{\partial t} + \xi \dfrac{\partial \xi}{\partial x} = \mathcal{A}[\rho, \xi] \,, \end{cases} \tag{6.2.9}$$

where $\mathcal{A}[\rho, \xi]$ models the mean acceleration and the square brackets indicate that it may be a functional of its arguments. For instance, some authors assume that $\mathcal{A}[\rho, \xi]$ is a suitable function of ρ and ξ, and of their space derivative.

The normalization condition (6.2.8), obtained by setting equal to one the coefficient of the second differential term, is equivalent to requiring that the characteristic time T is equal to the time needed to cover the whole length of the road at the maximum velocity.

The inlet and outlet of vehicles can be modeled by a source term $s = s(t, x)$, which may be referred to the dimensionless variables and localized in suitable space intervals $[x_h - H \, , \, x_h + H]$. The presence of a source term causes us to modify the right-hand side term of the first equation in (6.2.9), as well as the model of the acceleration term in the momentum conservation equation.

The framework above can be technically generalized. For instance, Aw and Rascle (2000) suggest, by heuristic arguments within a constructive replay to the critical analysis by Daganzo (1995), an alternative model stated in terms of two coupled conservation laws: the first one is the mass conservation law, and the second one corresponds to a suitable Riemann invariant. Specifically, the model is the following:

$$
\begin{cases}
\dfrac{\partial \rho}{\partial t} + \dfrac{\partial}{\partial x}(\rho \xi) = 0 \, , \\[2ex]
\dfrac{\partial}{\partial t}(\xi + p(\rho)) + \xi \dfrac{\partial}{\partial x}(\xi + p(\rho)) = 0 \, ,
\end{cases}
\qquad (6.2.10)
$$

where p is a heuristic expression of the pressure to be modeled by a suitable constitutive relation $p = p(\rho)$ which is taken as an increasing function of ρ. The simplest model is linear: $p = c\rho$.

Specific models can be designed referring to the above framework. For instance, **first-order models** are obtained by the first equation only with a closure $\xi = v_e[\rho]$ obtained by a phenomenological model describing the driver's psycho-mechanic adjustment of the velocity to a suitable local equilibrium velocity.

Analogously, **second-order models** are obtained by both equations (6.2.9) with the addition of a phenomenological relation describing the psycho-mechanic acceleration $\mathcal{A} = \mathcal{A}[\rho, \xi]$ on the vehicles.

This framework can be generalized to the case of roads with several lanes, where vehicles are allowed to move laterally from one lane to the other. The density over all lanes is a conserved quantity.

Traffic flow measurements are obtained by sophisticated devices which allow quite accurate measurements of macroscopic quantities such as number density, mean velocity, and flow. It is plain that the averaging process, which leads to macroscopic quantities from measurements on a system with finite degrees of freedom, introduces unavoidable fluctuations due not only to measurement errors, but also to the stochastic nature of the flow where deceleration and acceleration of vehicles are observed even in a steady flow. The order of magnitude of these fluctuations gives a rough idea of the accuracy of measurements.

Generally, experimental results refer to steady state conditions and to macroscopic quantities. Their representation, as documented in the books by Prigogine and Herman (1971), Leutzbach (1988), and Kerner (2004), reports the mean velocity or the flux versus the local density. Experiments may also give information on the spread of the measured quantities.

The mean dimensionless velocity $v_e = v_e(\rho)$ reaches its maximum value $\xi = 1$ for $\rho = 0$ and tends monotonically to zero for $\rho \to 1$, where the subscript e is used to identify the equilibrium conditions. Correspondingly, the flow $q_e = v_e \rho = q_e(\rho)$ starts from the value $q_e(0) = 0$, first increases, and then decreases to the value $q_e(1) = 0$. The above representations are often called, respectively, the **velocity diagram** and the **fundamental diagram**. The related modeling problem consists in looking for analytic expressions of v_e and q_e and, subsequently, in using these analytic expressions toward the derivation of hydrodynamical equations. For instance, the following formula:

$$v_e = (1 - \rho^{1+a})^{1+b}, \qquad q_e = \rho(1 - \rho^{1+a})^{1+b}, \qquad (6.2.11)$$

with $a, b \geq 0$, is due to Kühne and Rödinger, as reported in the review by Klar, Kühne, and Wegener (1996), where various alternative models are reported.

Relatively more recent experiments, such as those suggested by Bonzani and Mussone (2003), have shown that at low density vehicles generally keep the maximum velocity $v_e = 1$ until a critical value ρ_c of the density is reached. Then, for $\rho \geq \rho_c$, the velocity v_e starts decaying with increasing ρ. This critical value identifies a transition from free to congested flow. Moreover, note that when modeling flow conditions by analytic formulas, one should attempt to relate only one parameter to each specific phenomenon. This avoids the ambiguity that the same event may be described by different pairs of parameters. Therefore, as an alternative to (6.2.11), the following model is proposed in the same paper:

$$\begin{cases} \rho \leq \rho_c : & v_e = 1, \\ \\ \rho \geq \rho_c : & v_e = \exp\left\{ -\alpha \dfrac{\rho - \rho_c}{1 - \rho} \right\}. \end{cases} \qquad (6.2.12)$$

This model is characterized by two free parameters: ρ_c related to the transition from free to congested flow, and α related to the decay of the equilibrium velocity with increasing density. The model is able to capture, with the same number of parameters, additional phenomena with respect to the one reported in Eq. (6.2.11).

The analysis of experimental data given in the above-cited paper suggests the following ranges for the admissible domains for the parameters:

$$\rho_c \in D_c = [0, 0.2], \qquad \alpha \in D_\alpha = [1, 2.5],$$

depending on hour, weather, seasonal conditions, etc. The whole set of outer conditions is called, in what follows, the *environmental conditions*.

Some authors use the above empirical data as an input to derive models, while, at least in principle, a careful modeling of microscopic interactions in the case of kinetic models should provide the above result.

6.2.3 Representation by Kinetic Theory Methods

The state of the whole system is defined, for each lane, by the statistical distribution of position and velocity of the vehicles. Specifically consider, for a one-lane road, the following distribution over the dimensionless microscopic state

$$f = f(t, x, v), \tag{6.2.13}$$

where now $f(t, x, v)dx\, dv$ gives the number of vehicles which, at time t, are in the phase space domain $[x, x+dx] \times [v, v+dv]$. The distribution function f can be normalized with respect to n_M so that all derived variables can be given in a dimensionless form.

Macroscopic observable quantities can be obtained, under suitable integrability assumptions, by moments of the above distribution function. In particular, the *dimensionless local density* is given by

$$\rho(t, x) = \int_0^{1+\mu} f(t, x, v)\, dv. \tag{6.2.14}$$

The *total number of vehicles* at the time t is given as

$$N(t) = \int_0^1 \int_0^{1+\mu} f(t, x, v)\, dv\, dx. \tag{6.2.15}$$

In the same way, the local **mean velocity** can be computed as follows:

$$\xi(t, x) = \frac{1}{\rho(t, x)} \int_0^{1+\mu} v\, f(t, x, v)\, dv. \tag{6.2.16}$$

Similarly, the local **speed variance** is given by

$$\sigma(t,x) = \frac{1}{\rho(t,x)} \int_0^{1+\mu} [v - \xi(t,x)]^2 \, f(t,x,v) \, dv \,. \tag{6.2.17}$$

The above macroscopic quantities can also be related to the **flow** and the **speed pressure**, respectively q and p, as follows:

$$\xi(t,x) = \frac{q(t,x)}{\rho(t,x)} \,, \tag{6.2.18}$$

where

$$q(t,x) = \int_0^{1+\mu} v f(t,x,v) \, dv \,, \tag{6.2.19}$$

and

$$p(t,x) = \sigma(t,x) \, \rho(t,x) \,, \tag{6.2.20}$$

where

$$p(t,x) = \int_0^{1+\mu} [v - \xi(t,x)]^2 \, f(t,x,v) \, dv \,. \tag{6.2.21}$$

Additional technical notation is needed in the case of multilane flows, where the superscript j to the distribution function identifies the lane. When useful, one can obtain quantities which are averaged over all lanes. For instance, the lane-averaged mean velocity is

$$\xi(t,x) = \frac{1}{R} \sum_{j=1}^R \xi^j[v](t,x) \,, \tag{6.2.22}$$

where

$$\xi^j[v](t,x) = \frac{1}{\rho^j(t,x)} \int_0^{1+\mu} v f^j(t,x,v) \, dv \,. \tag{6.2.23}$$

As known, a continuous representation requires dealing with a large number of particles. For this reason some recent papers have proposed discrete velocity models such that the velocity variable can attain only a finite number of values belonging to a set of admissible velocities.

The mathematical structure depends both on the method of modeling microscopic interactions and on the discretization of the velocity space. Different structures can be used, still in agreement with those of Chapters 2–4, according to a detailed modeling of microscopic interactions. Detailed calculations are reported in Sections 6.4 and 6.5.

6.2.4 Further Analysis on the Selection of the Representation Scale

The various representation schemes given in the preceding subsections have been referred to the usual three scales from the microscopic to the macroscopic, through the statistical representation. However, some simple reasonings show that none of the above representations can be regarded as effectively consistent with the complex system we are dealing with.

The analysis offered by Daganzo (1995) clearly stresses these considerations. The following ideas are extracted from this paper:

i) *Shock waves and particle flows in fluid dynamics involve thousands of particles, while only a few vehicles are involved by traffic jams.*

ii) *A vehicle is not a particle but a system linking driver and mechanics, so that one has to take into account the reaction of the driver, who may be aggressive, timid, prompt etc. This criticism also applies to kinetic type models.*

iii) *Increasing the complexity of the model increases the number of parameters to be identified.*

It is plain that the number of vehicles in traffic flow conditions is not, even in congested traffic, large enough to ensure the validity of continuity of matter which is necessary to approach a hydrodynamic-type modeling. In addition, it is not even sufficient for the statistical description of the kinetic theory. Therefore, the distribution function cannot be regarded as continuous with respect to the variables describing the microscopic state.

This reasoning supports the idea that modeling is consistent only at the microscopic scale. However, high complexity problems are generated due to the large number of differential equations necessary to describe the system. In addition, one has to deal with a technically difficult averaging process to obtain information at the macroscopic level. Suitable filters can be applied, however accuracy of computing is hard, perhaps even impossible, to prove.

Some remarkable sources of complexity can be summarized, among several, in view of a modeling approach that is likely to deal efficiently with them.

- The system is definitely discrete, i.e., with finite degrees of freedom. However it is necessary, for practical purposes, that the model allows the computation of macroscopic quantities.

- The flow is not continuous, hence hydrodynamic models should not be used, considering that it is difficult, or even impossible, to evaluate the entity of approximation, say falsification, these models induce with respect to physical reality.

- The number of particles is not large enough to allow the use of continuous distribution functions within the framework of the mathematical kinetic theory. Moreover, interactions are not localized as in the case of classical particles, because drivers adapt the dynamics of the vehicle to the flow conditions in front of them.

- Vehicles cannot be regarded as classical particles, but as active particles due to the presence of the driver, who modifies the dynamics of the vehicle following specific strategies.

The last source of complexity appears to be particularly relevant, although the literature in the field has not yet dealt carefully with this aspect of the modeling approach. Models take somehow into account the fact that the dynamics of a vehicle is related to the presence of the driver. However, the modeling is generally based on the idea that all driver-vehicles behave in the same manner. In other words, Daganzo's criticism can be applied to almost all models in the literature.

The modeling of the features above can possibly be achieved by inserting the activity variable into the microscopic state, while the distribution over such a variable may depend on the flow conditions. For instance, the differences among various driver-vehicles are negligible in traffic jams, while they are relevant in free flow conditions.

As already mentioned, the existing literature discusses these topics only at a preliminary level. The sections which follow provide some speculations on new ideas addressed to tackle these problems.

Finally, let us remark that when one compares the prediction of the model with empirical data, one has to consider the fact that experimental data are collected, and analytically described, in conditions which are stationary in time and uniform in space. However, the prediction of models should refer to unsteady flow conditions, generally far from the stationary uniform flow.

6.3 A Survey of Continuous Kinetic Traffic Models

This section provides a survey of some traffic flow models designed within the framework of mathematical kinetic theory. This presentation aims at providing a background to some of the developments in the sections which follow. The literature in the field is vast and includes several interesting contributions, although they are not yet based on the methods of kinetic theory for active particles. Specifically, although interactions do not follow the rules of classical mechanics, their modeling is the same for all vehicles without considering the role of the activity variable.

Completeness is not claimed in this concise review; the aim is simply to show, by a few examples, how different methodological approaches can be proposed in view of their development by looking at the driver-vehicle subsystem as an active particle.

As already mentioned, modeling traffic flow in a Boltzmann-like manner was initiated by Prigogine and Herman (1971). After the above pioneer approach, several authors developed interesting improvements of their model.

The model is derived assuming that the driver is willing to adjust its velocity, either by increasing or by decreasing it, towards a certain desired velocity distribution. In addition, velocity may change due also to the interaction with the leading vehicle. In both cases, the rate of change depends on the density of vehicles.

The flow is assumed to be one dimensional, and each vehicle is modeled as a point, i.e., the length of each vehicle is negligible with respect to the length of the road, although a maximal density ρ_M is considered. The state of each vehicle at time t is defined by its position x and velocity v. The state of the system is described by the distribution function $f = f(t, x, v)$. The evolution of f is ruled by a balance equation, generated by vehicle interactions, according to the scheme

$$\frac{\partial f}{\partial t} + v\,\frac{\partial f}{\partial x} = J[f] = J_r[f] + J_i[f]\,, \qquad (6.3.1)$$

where J_r is the relaxation term, which accounts for the speed change towards a certain **program** of velocities independent of local concentration, and J_i is the term due to the **slowing down** interaction between vehicles.

The term J_r is modeled assuming that drivers have a program in terms of a desired velocity described by the distribution $f_e = f_e(x, v)$. This is called the **desired-velocity distribution function**. Moreover, the drivers also desires to this velocity distribution within a certain relaxation time T_r, related to the normalized density and equal for all drivers.

Prigogine's relaxation term is the following:

$$J_r[f](t, x, v) = \frac{1}{T_r[f]}\,(f_e(x, v) - f(t, x, v))\,, \qquad (6.3.2)$$

where T_r depends on the density

$$T_r[f](t, x) = \tau\,\frac{\rho[f](t, x)}{1 - \rho[f](t, x)}\,, \qquad (6.3.3)$$

where τ is a constant.

The term J_i is due to the interaction between a **test** vehicle and its leading **field** vehicle. It accounts for the changes in $f(t, x, v)$ caused by braking of the test vehicle due to an **interaction** with a field vehicle, and it refers to the braking when the test vehicle has velocity $v < w$, and an acceleration term when the field vehicle has velocity $w < v$. Moreover, J_i is proportional to the probability P that the fast car may pass the slower one, and which may be related to the normalized density assumed to be equal for all drivers.

Prigogine's interaction term is defined by

$$J_i[f](t, x, v) = (1 - P[f])f(t, x, v) \int_0^{1+\mu} (w - v)f(t, x, w)\, dw \,, \qquad (6.3.4)$$

where

$$P[f](t, x) = 1 - \rho[f](t, x) \,. \qquad (6.3.5)$$

According to the above assumptions, the explicit expression of the mathematical model is derived by the usual balance in the elementary volume of the phase space under the usual assumption of probability independence of the interacting pairs:

$$\frac{\partial f}{\partial t} + v\,\frac{\partial f}{\partial x} = \frac{1}{\tau}\,\frac{1 - \rho[f](t, x)}{\rho[f](t, x)}\,\left(f_e(x, v) - f(t, x, v)\right)$$

$$+\rho[f](t, x)\,f(t, x, v) \int_0^{1+\mu} (w - v)f(t, x, w)\, dw \,. \qquad (6.3.6)$$

The first relevant modification of the model was given by Paveri Fontana (1975), who criticizes the relaxation term by showing that it has some unacceptable consequences; for instance, it becomes meaningless for low densities. To overcome such a problem, the desired velocity v^* is assumed to be an independent variable of the problem, and a generalized one-vehicle distribution function $g = g(t, x, v; v^*)$ is introduced to describe the distribution of vehicles at (t, x) with speed v and desired speed v^*. Hence the distribution f_e that concerns the desired speed and the distribution f that concerns the actual speed are given by

$$f_e(t, x, v^*) = \int_0^{1+\mu} g(t, x, v; v^*)\, dv \,, \qquad (6.3.7)$$

and

$$f(t, x, v) = \int_0^{1+\mu} g(t, x, v; v^*)\, dv^* \,. \qquad (6.3.8)$$

The evolution equation, which now refers to the generalized distribution function g, is again determined by equating the transport term on g to the sum of the interaction term and relaxation term:

$$\frac{\partial g}{\partial t} + v\frac{\partial g}{\partial x} \;=\; J_P[g] \;=\; J_i[g] + J_r[g] \,. \qquad (6.3.9)$$

The interaction term has the same structure as in the previous case, however, the operators apply to g, and the passing probability P is assumed to also be a function of a certain critical density ρ_c. Paveri Fontana's interaction term is defined by

$$J_i[g](t, x, v; v^*) = (1 - P[f])f(t, x, v) \int_v^{1+\mu} (w - v)g(t, x, w; v^*) \, dw$$

$$- (1 - P[f])g(t, x, v; v^*) \int_0^v (v - w)f(t, x, w) \, dw , \qquad (6.3.10)$$

where

$$P[f](t, x) = (1 - \rho[f](t, x))H(1 - \rho[f](t, x)) , \qquad (6.3.11)$$

and H is the Heaviside function.

The relaxation towards a certain program of velocities is related to the vehicles' acceleration. This is taken into account by means of a relaxation time T_r that is seen as a function of the passing probability P, and hence of the density:

$$J_r[g](t, x, v; v^*) = -\frac{\partial}{\partial v} \left(\frac{v^* - v}{T_r[f]} \, g(t, x, v; v^*) \right) , \qquad (6.3.12)$$

where

$$T_r[f](t, x) = \tau \, \frac{1 - P[f](t, x)}{P[f](t, x)} . \qquad (6.3.13)$$

The mathematical model is obtained by substituting the preceding equations into (6.3.9).

Recently, various authors have developed kinetic models based upon a detailed microscopic description of the pair interactions with the aim of extending the two pioneering models we have just recalled. Here we briefly outline some of the efforts recently made as examples of possible developments of Prigogine's and Paveri Fontana's theory.

The collision operator is modeled, on a mechanical pairwise interaction basis, by analyzing each driver's short range reactions to neighborhood vehicles rather than by interpreting his overall behavior. Interactions are strictly pairwise since the test vehicle only reacts to what happens in the vehicle traveling in front of it. This makes the model more similar to the Boltzmann kinetic scattering equation, and in some sense closer to the follow-the-leader point of view.

Models stemming from the microscopic description of pair interactions have the advantages that comparisons with experimental data (and organization of suitable experiments) can be made only in relati to the mentioned microscopic behaviors. Moreover, stationary solutions, which are of great relevance in the analysis of traffic flow, are direct predictions of the model. The new approach was introduced and developed by various authors, e.g., Nelson (1995), and Klar and Wegener (1997), (2000), who were able to exploit the advantages of a modeling based upon describing the short range interactions.

However, modeling microscopic interactions is not a simple task as it requires a detailed analysis of vehicle dynamics and driver's reactions, together with the organization of specifically related experiments. The problem consists in finding suitable expressions for the post-interaction velocities v' and w' which, in the microscopic modeling, are directly related to the pre-interaction ones: v and w. Furthermore, if high densities must be taken into account, modifications of the interaction frequency should be included in a way similar to that followed in deriving Enskog equations.

Nelson (1995) develops a model, which he calls **a traffic flow caricature**, which is based on some technically restrictive assumptions on the dynamics allowed to each driver. Some of them are:

i) Zero passing probability;

ii) A unique value for the desired velocity which coincides with the highest possible speed v_M;

iii) A unique minimal headway distance;

iv) A particular kind of vehicular chaos;

v) Instantaneous changes of speed and reactions to any circumstances.

These last two assumptions imply two different time scales: an immediate time scale and an evolution time scale. The author obtains the interesting result of a bimodal equilibrium distribution, centered at the desired (maximal) speed $v_M = 1$ and at the zero speed, which is somewhat surprising. As a matter of fact, this bimodal distribution is related to the time scale of immediate reactions.

Moreover, Nelson is the first author who treats the speeding-up event in essentially the same way as the slowing-down event, proposing a **table of the truth** of outgoing velocities in response to each possible incoming circumstance. In other words, a transition probability density $\psi(v, v')$ is a *priori* sketched. Hence, he doesn't need to introduce any relaxation term such as those of the Prigogines and Paveri Fontanas models.

In more detail, Nelson's model is based on the following additional assumption:

vi) The set of possible values for the test vehicle's outgoing velocity after a change-in-speed event, or **interaction**, is restricted to

$$\{0, v, v_H, v_M = 1\},$$

where v_H denotes the leading vehicle velocity.

Fundamental to any interaction process, because it triggers the occurrence of the change-in-speed event, is the **minimal headway** $k(v)$. The headway distance k is assumed to be a strictly increasing function of just one velocity, mostly the leading vehicle speed. A suitable **headway probability density** $p(h|t, x, v)$ is introduced to account for the probability that the headway distance may be less than a certain h. The transition probability ψ is constructed depending upon the various possible values of h with respect to the desired value $k(v)$.

In addition, the following technical assumptions are stated:

$$p(h|t, x, v) = p(h|t, x), \tag{6.3.14}$$

$$p(h|t, x) = \begin{cases} 1 - \exp(-\rho(t, x)(k - h(0))) & \text{if } h > k(0), \\ 0 & \text{if } h \leq k(0), \end{cases} \tag{6.3.15}$$

$$g(h, v_H; t, x, v) = \frac{f(t, x, v_H)}{\rho(t, x)} \frac{dp}{dh}(h|t, x), \tag{6.3.16}$$

where $g(h, v_H; t, x, v)$ denotes the probability density that a test vehicle has a leading vehicle at headway h with velocity v_H. The model proposed by Nelson follows by technical calculations which make use of the above table of interactions.

Starting from Nelson's model, to release some of its restrictive assumptions and to gain a more general and flexible structure, Klar and Wegener (2000) make use essentially of the same mathematical structure of gain and loss term that is familiar in the Boltzmann theory. A detailed description of microscopic interaction leads to models which are able to describe, in the case of stationary solutions, a distribution over the velocity of the type observed experimentally. Specifically, the distribution tends to zero for $v \to 0$ and for large values of v; moreover, the distribution tends to a shape similar to a delta function when the density tends to its maximum value: $\rho \to 1$. This is a remarkable result to be carefully considered when developing new models.

The preceding concise survey shows that models refer to different mathematical structures, but still without including the activity variable in the description, at the microscopic level, of the driver-vehicle system. The frameworks can be classified as typical of phenomenological models, and local binary interaction models. It is worth analyzing, referring to Chapters 2–4, the structures above in view of some conceivable generalizations which include averaged binary and mean field interactions.

Let us first consider **phenomenological models** with a structure as follows:

$$\frac{\partial f}{\partial t} + v \frac{\partial f}{\partial x} + F(t, x) \frac{\partial f}{\partial v} = Q[f; \rho], \tag{6.3.17}$$

where $f = f(t, x, v)$ is the distribution function of the **test** vehicle, F is the acceleration applied to it by the outer environment, and Q is a suitable function, derived on a phenomenological basis, of f, which may be parameterized by local gross quantities.

A simple way to model the term Q consists in describing a trend to equilibrium analogous to the BGK model in kinetic theory:

$$Q = c_r(\rho)\left(f_e(v; \rho) - f(t, x, v)\right), \qquad (6.3.18)$$

where the rate of convergence is c_r and f_e denotes the equilibrium distribution function, which may be assumed to depend on the local density.

Models generally assume $F = 0$. However, acceleration terms may be imposed by the outer environment, e.g., signaling to accelerate or decelerate. Still, a question to be naturally posed refers to the possibility that a vehicle may be subject to an acceleration due to the vehicles ahead similar to the one characterizing hydrodynamical models corresponding to conservation of mass and momentum.

Localized binary interaction models are based on microscopic modeling which assumes binary interaction between the test and the field vehicles localized in the point x of the field vehicle. Interactions, similarly to the Enskog equation, can be localized at a fixed distance d from the test vehicles on its front. Moreover, it is assumed, similarly to the Boltzmann equation, factorization of the joint probability related to the microscopic state of the two interacting vehicles.

For both types of interactions, the formal structure of the evolution equation is as follows:

$$\frac{\partial f}{\partial t} + v \frac{\partial f}{\partial x} = J[f], \qquad (6.3.19)$$

where J can be written, according to Chapters 2–4, as the difference between the inflow (***gain***) and outflow (***loss***) of vehicles in the control volume of the phase space.

An alternative to this approach can be developed by modeling ***averaged binary interactions***. This approach consists in assuming that the driver of the test vehicle in x has a visibility zone $[x, x + c_v]$ on the front, while interactions are weighted, within the above zone, by a suitable weight function $\varphi = \varphi(y)$, with $y \in [x, x + c_v]$ such that

$$y \uparrow \quad \Rightarrow \quad \varphi \downarrow, \quad \text{and} \quad \int_x^{x+c_v} \varphi(y)\, dy = 1. \qquad (6.3.20)$$

The modeling is simply given by a technical generalization of (6.3.19) obtained simply by weighting the interaction term over the variable $y \in [x, x + c_v]$.

Drivers organize the dynamics of the vehicle according to mean field stimuli in the visibility zone. In this case the mathematical structure can be modified as follows:

$$\frac{\partial f}{\partial t} + v\frac{\partial f}{\partial x} + \frac{\partial}{\partial v}\left(fA[f]\right) = 0\,, \qquad (6.3.21)$$

where $A[f]$ is the mean field acceleration applied by the field vehicles in the visibility zone of the test vehicle.

All of these structures can be used towards the derivation of specific models. A suitable combination of them can also be exploited.

6.4 Discrete Velocity Models

Different mathematical structures can be used, as we have seen in the preceding section, to model traffic flow phenomena, yet one cannot precisely identify a general method to select as the optimal one. A further problem which has not yet been fully analyzed is the validity of the continuous representation of the distribution function. This assumption, in kinetic theory for classical particles, is justified by the fact that a fluid, even in the case of a rarefied gas, is constituted by millions of particles. However, the number of vehicles on a road is limited to a very short number of entities. This matter has already been critically analyzed in Section 6.2.

An attempt to take this issue into account is offered by discrete velocity models, which correspond to a fluid where particles can attain only a finite number of velocities. The mathematical kinetic theory for discrete velocity fluids was introduced by Gatignol (1975) to reduce the structural complexity of the full Boltzmann equation. The literature in the field is documented in the survey paper by Platkowski and Illner (1985) and in the collection of surveys edited by Bellomo and Gatignol (2003).

Discretization of the velocity variable in traffic flow modeling should not be regarded as a method to reduce complexity, but as a means to solve, at least partially, the problem of the difficulty of a continuous representation of the distribution function. Indeed, models refer to groups of vehicles within cells of the velocity space. This mathematical approach was recently introduced by Coscia, Delitala, and Frasca (2007), who used a grid depending on the local density. The contents of this section refer to this paper and are developed through three subsections which deal with the following topics:

i) Mathematical structures for discrete velocity modeling;

ii) Some ideas on discretization of the whole phase space;

iii) Concise description of the CDF model;

6.4.1 Mathematical Structures

Let us consider *discrete velocity models*, where the velocity variable belongs to the following set:

$$I_v = \{v_1 = 0, \dots, v_i, \dots, v_n = 1\}, \tag{6.4.1}$$

where velocities have been divided by the maximal admitted one: V_ℓ.

The corresponding discrete representation is obtained by linking the discrete distribution functions to each v_i:

$$f_i = f_i(t, x) : \quad \mathbb{R}_+ \times [0, 1] \to \mathbb{R}_+, \tag{6.4.2}$$

for $i = 1, \dots, n$.

This discrete velocity approach naturally implies that the number of vehicles with a velocity larger than V_ℓ can be disregarded. In other words, it is technically assumed that the presence of vehicles, with velocity much larger than the maximum mean velocity corresponding to the given density, is neglected.

The above-cited paper by Coscia, Delitala, and Frasca (2007) prefers considering the interval $[0, 1 + \mu]$, to identify the dimensionless velocity domain, where $\mu > 0$ is a parameter accounting for the fact that under very light traffic conditions vehicles may attain velocities greater than v_M. However, in a discrete-velocity framework such a detail is not very relevant, since vehicles are grouped and classified on the basis of velocity classes $\{v_i\}_{i=1}^n$. Those which possibly travel at speeds higher than v_M are simply included in the extreme class v_n.

We stress again, that discretization of the velocity domain, which has been introduced above, is not a mere mathematical procedure, but represents a possible manner to take into account, at least partially, the strongly granular nature of traffic. Regarding this, we observe that vehicles traveling along a road do not span the whole set of admissible velocities; rather, they tend to move in clusters, which can be identified and distinguished from each other by a discrete set of velocity values.

According to the method above, the following gross quantities are obtained by weighted sums:

$$\rho(t, x) = \sum_{i=1}^{n} f_i(t, x), \tag{6.4.3}$$

and

$$q(t,x) = \sum_{i=1}^{n} v_i f_i(t,x) , \qquad \xi(t,x) = \frac{q(t,x)}{\rho(t,x)} . \qquad (6.4.4)$$

We also define, following the gas kinetic theory, two quantities that can give important information about the vehicle flux: the speed variance and the H functional. Respectively,

$$\sigma(t,x) = \frac{1}{u(t,x)} \sum_{i=1}^{n} (v_i - \xi(t,x))^2 f_i(t,x) , \qquad (6.4.5)$$

and

$$H(t,x) = \sum_{i=1}^{n} f_i(t,x) \log f_i(t,x) . \qquad (6.4.6)$$

In general, the model consists of a set of evolution equations for the densities f_i, where the mathematical structure of the evolution equations in the spatially homogeneous case can be written as follows:

$$\frac{df_i}{dt} = \sum_{h=1}^{n} \sum_{k=1}^{n} \eta_{hk} A_{hk}^i f_h f_k - f_i \sum_{k=1}^{n} \eta_{ik} f_k , \qquad (6.4.7)$$

for $i = 1, \ldots, n$, where:

• η_{hk} is the encounter rate (number of interactions per unit time) of vehicles with velocities v_h and v_k;

• A_{hk}^i is the probability density that a vehicle with velocity v_h, the **test** or **candidate** vehicle, reaches the velocity v_i after the interaction with the vehicle with velocity v_k, the **field** vehicle.

Reversibility, which is a typical feature of classical particles, is not claimed here. Indeed, the output of the interaction depends on the ability of drivers to organize the dynamics of the vehicle.

The derivation of the model, consistent with the above framework, means modeling the encounter rates and the transition probability densities according to the specific phenomenological behavior of the system.

Modeling the **encounter rate** is a matter of mechanical calculations. This quantity depends on the relative velocity though a dimensional parameter γ:

$$\eta_{hk} = \gamma |v_h - v_k| , \qquad (6.4.8)$$

Where for technical calculations $\gamma = 1$ is chosen.

Modeling the terms A^i_{hk} requires a detailed analysis of microscopic interactions. The minimal structure to deal with the modeling of the spatially inhomogeneous case is as follows:

$$\left(\frac{\partial f_i}{\partial t} + v_i \frac{\partial f_i}{\partial x}\right)(t, x) = J_i[\mathbf{f}]$$

$$= \sum_{h=1}^{n} \sum_{k=1}^{n} \gamma |v_h - v_k| A^i_{hk} f_h(t, x) f_k(t, x)$$

$$- f_i(t, x) \sum_{k=1}^{n} \gamma |v_i - v_k| f_k(t, x), \tag{6.4.9}$$

where the transition probability densities A^i_{hk} refer to the probability that a candidate vehicle with velocity v_h will reach the velocity v_i after an interaction with a field vehicle with velocity v_k, and γ is a dimensional constant.

The above structure is valid if the velocity discretization is defined by a constant grid and if the terms A^i_{hk} are constant quantities. Simply, it is required that these terms have the structure of a discrete probability density:

$$\sum_{i=1}^{n} A^i_{hk} = 1, \qquad \forall h, k = 1, \ldots, n. \tag{6.4.10}$$

6.4.2 On the Discretization of the Whole Phase Space

An additional framework can be designed by a double discretization that also includes the space variable. Let us consider an equally spaced grid in the space variable of the type

$$I_x = \{x_1 = 0, \ldots, x_j, \ldots, x_m = 1\}, \tag{6.4.11}$$

where the space interval $d_j = x_j - x_{j-1}$, which identifies volume cells, should be less than the visibility zone. Moreover, let

$$f_{ij}(t) = f(t, v_i, x_j), \tag{6.4.12}$$

so that the model consists of an evolution equation for the discrete distribution function.

The derivation of the model can be developed by approximating the space derivative by a conservative scheme using the values of f in the nodes x_j and x_{j+1}. Similarly, the term J_i can be properly weighted by its value in the cells in the front of x_j.

Therefore, the formal structure is as follows:

$$\frac{df_{ij}}{dt} + v_i \mathcal{D}_{ij}[f_{ij}, \dots, f_{i,j+r}] = \sum_{p=j}^{p=j+r} J_{ip}[\mathbf{f}]\bar{w}_p \,, \qquad (6.4.13)$$

where the weights are such that

$$\sum_{p=j}^{p=j+r} \bar{w}_p = 1 \,. \qquad (6.4.14)$$

Further detailed calculations must be developed to obtain specific models.

6.4.3 The CDF Model

The derivation of specific models, consistent with the above framework, means modeling the transition probability densities according to the specific phenomenological behavior of the system.

A discrete velocity kinetic model for vehicular traffic has also been developed in the paper by Coscia, Delitala, and Frasca (2007). The modeling of the transition probability densities is based on a discretization where the admissible values v_i of the velocity depend on the average local traffic conditions via the macroscopic density $\rho = \rho(t, x)$. Specifically, the grid I_v is conceived to have a variable step Δv, which tends to zero for high vehicle concentrations (**adaptive velocity grid**).

In detail, the following discretization is adopted:

$$I_v = \{v_1 = 0, \dots, v_i, \dots, v_n = v_e(\rho), \dots, v_{2n-1} = 2v_e(\rho)\} \,, \qquad (6.4.15)$$

where $i = 1, \dots, 2n - 1$ with

$$v_i = \frac{i-1}{n-1} v_e(\rho) \,,$$

and where v_e is the velocity under uniform steady state flow conditions obtained by experimental data and approximated as indicated in Section 6.2

The modeling of the terms A_{hk}^i, which define the **table of games**, requires a mathematical interpretation of the microscopic phenomenology of the system. The specific assumptions which generate the model are the following:

- Interactions modify the velocity of the test and field vehicles only if v_h and v_k are sufficiently close:

$$A_{hk}^i = 0 \quad \text{if} \quad |h - k| > 1; \qquad i = 1, \dots, 2n - 1 \,.$$

– The test vehicle can modify its velocity only by jumping to a neighboring velocity value:

$$A^i_{hk} = 0 \quad \text{if} \quad |i - h| > 1 \quad \text{for} \quad k = h - 1, \quad h, \quad h + 1.$$

According to the above-stated assumptions, only the nonnull terms of matrices A^i_{hk}:

$$A^{i=h-1}_{h\,h-1}, \ A^{i=h}_{h\,h-1}, \ A^{i=h+1}_{h\,h-1}, \ A^{i=h+1}_{h\,h+1}, \ A^{i=h}_{h\,h+1}, \ A^{i=h-1}_{h\,h+1},$$

need to be considered.

– When $h < n$, the test vehicle is slower than the equilibrium velocity and the driver has a tendency to increase his speed when he/she interacts with a fast vehicle, while no changes occurs if he/she interacts with a slower vehicle. The opposite trend is expected when a test vehicle that is faster than the equilibrium velocity ($h > n$) interacts with a slower vehicle.

– The following **table of games** provides a mathematical interpretation of the phenomenology of the system:

$$\text{if } h \leq n \text{ and } k = h + 1: \quad A^{h+1}_{h\,h+1} = \varepsilon_a, \quad A^h_{h\,h+1} = 1 - \varepsilon_a,$$

$$\text{if } h \geq n \text{ and } k = h - 1: \quad A^{h-1}_{h\,h-1} = \varepsilon_d, \quad A^h_{h\,h-1} = 1 - \varepsilon_d,$$

$$\text{if} |h - k| > 1 \quad \text{or} \quad k = h: \quad A^h_{hk} = 1, \quad \text{otherwise} \quad A^i_{hk} = 0,$$

for $(i, h = 1, \ldots, 2n - 1; \ k = h - 1, h, h + 1)$.

The accelerating probability ε_a and the decelerating probability ε_d represent the reactivity of vehicles to different events and in principle may assume different values if one of the events is more effective; e.g., braking usually has a shorter reaction time with respect to acceleration. It is useful to rewrite them as $\varepsilon_a = \varepsilon$ and $\varepsilon_b = \nu \varepsilon$ to identify a basic reaction probability ε and a measure of its asymmetry ν.

The preceding assumptions model the fact that slow vehicles have a trend to reach the mean velocity when *motivated* by fast vehicles, while the opposite behavior is observed for fast vehicles, which are *motivated* to decelerate by the presence of slow vehicles. Interactions are supposed to be effective only when the velocity distance is not too large.

The model is identified by five parameters only: μ, γ, n, ε, and ν.

• μ is a parameter that identifies the maximum velocity of a single vehicle and can be directly identified from experimental data.

• γ is a dimensional parameter which is included in the time variable and set equal to one.

- n defines the number of nodes. In particular, n defines a velocity range of sensitivity, in which interactions are effective.

- ε corresponds to the tendency of the driver to modify his/her velocity when interacting with a vehicle with a velocity in the above-defined interacting velocity range.

- ν takes into account the different tendency in accelerating or braking of the driver. Actually it is shown, in the above-cited paper, that the choice of ν heavily affects the shape of the equilibrium distribution.

The mathematical structure of the model is a nonhomogeneous system of hyperbolic first-order equations with a quadratic right-hand-side term in the unknowns. Moreover, since the velocity v_i depends on ρ, namely on f_i, the left term is nonlinear too. Interactions between vehicles are modeled both explicitly by the table of games and implicitly by the empirical equation $v_e(\rho)$, which instantaneously adapts the velocity discretization to the density.

Various simulations are given in the above-cited paper that show how the model can provide a qualitative description of various phenomena which are observed in physical reality. An interesting feature of the model is the grid with variable size, that naturally adapts the intensity of the interaction to the local density: the higher the density, the lower the intensity of the interactions.

Still the model needs suitable assumptions on the behavior of v_e versus ρ, because the development described in Section 6.5 is only based on a detailed assumption of microscopic interaction, while steady flow conditions are described by the model itself.

6.5 On the Model by Delitala and Tosin

The mathematical approach proposed by Delitala and Tosin (2007) substantially differs from that one of Section 6.4 although it still uses, according to the same motivations given above, discrete velocity methods. However, interactions are not localized, but averaged in the visibility zone of the driver.

Technically, a different discretization is used by dividing the velocity variable into equally spaced intervals, namely a **fixed velocity grid** is adopted and is identified by v_i for $i = 1, \ldots, n$. The overall state of the system, as in Section 6.4, is described by the distribution functions

$$f_i(t, x), \qquad i = 1, \ldots, n. \tag{6.5.1}$$

The adopted discretization method is such that a vehicle can attain a set of velocities which is not influenced by local traffic conditions:

$$v_i = (i-1)\Delta v, \qquad i = 1, \ldots, n \tag{6.5.2}$$

over the velocity domain $D_v = [0, 1]$, where

$$\Delta v = \frac{1}{n-1}.$$

The main feature of the model is that interactions are distributed over a (dimensionless) characteristic length $\xi > 0$, which can be interpreted as the **visibility length** of the driver: a vehicle located at a point $x \in D_x$ is supposed to be affected on average by all vehicles comprised within a certain **visibility zone**, which in this context is identified with the interval $J_\xi(x) = [x, x+\xi]$.

In addition, interactions are weighted by means of a proper function w, and occur with more or less frequency according to the free space locally available in the visibility zone. Considering that a vehicle is essentially an anisotropic particle, in the sense that it reacts mainly to frontal stimuli instead of rear ones, the assumption that the visibility zone does not include any stretch behind the vehicles appears to be consistent with the problem at hand.

The mathematical structure to generate specific models is summarized by the following equation:

$$\frac{\partial f_i}{\partial t} + v_i \frac{\partial f_i}{\partial x} = J_i[\mathbf{f}]$$

$$= \sum_{h=1}^{n} \sum_{k=1}^{n} \int_x^{x+\xi} \eta[\mathbf{f}](t, y) A_{hk}^i[\mathbf{f}](t, y) f_h(t, x) f_k(t, y) w(x, y) \, dy$$

$$- f_i(t, x) \sum_{h=1}^{n} \int_x^{x+\xi} \eta[\mathbf{f}](t, y) f_h(t, y) w(x, y) \, dy, \tag{6.5.3}$$

where $\eta[\mathbf{f}]$ is the **interaction rate**, which gives the number of interactions per unit time among the vehicles; and $A_{hk}^i[\mathbf{f}]$ defines the **table of games**, which models the microscopic interactions among vehicles by giving the probability that a vehicle with speed v_h will adjust its velocity to v_i after an interaction with a vehicle traveling at speed v_k.

Delitala and Tosin assume that both terms depend on the local density ρ with the additional requirement

$$A_{hk}^i[\rho] \geq 0, \quad \sum_{i=1}^{n} A_{hk}^i[\rho] = 1, \quad \forall h, k = 1, \ldots, n, \quad \text{for all} \quad \rho. \tag{6.5.4}$$

Moreover, $w(x, y)$ represents the function weighting the interactions over the visibility zone and is required to satisfy

$$w(x, y) \geq 0, \qquad \int_x^{x+\xi} w(x, y) \, dy = 1 \,. \qquad (6.5.5)$$

Equation (6.5.3) shows the structure of a system of integro-differential equations with a hyperbolic linear advection term. Borrowing some ideas from the Enskog kinetic theory of dense gases (see Ferziger and Kaper (1972), Bellomo, Lachowicz, Polewczak, and Toscani (1991)), the rate η of the interactions among the vehicles is assumed to be inversely proportional to the density ρ:

$$\eta[\rho] \simeq \frac{1}{1 - \rho} \qquad (6.5.6)$$

for $\rho \in [0, 1)$.

Note that this function is monotonically increasing with respect to ρ in the interval $[0, 1)$, which implies that the local interaction rate becomes higher and higher as the density increases toward its limit threshold fixed by the road capacity.

The **table of games** $A_{hk}^i[\rho]$ models the microscopic interactions among the vehicles, yielding the probability that the candidate vehicle will change its state class from h to that of the test vehicle i as a result of an interaction with a field vehicle in the state class k.

In modeling microscopic interactions, a key role is played by the density ρ, intended as an indicator of the macroscopic local conditions of the traffic. Moreover, the model considers an additional factor which strongly affects the flux of vehicles, namely the road conditions: bumpy roads make people drive more carefully, maintaining slow speeds and avoiding accelerations and passing, while smooth roads usually offer more opportunities for maneuvering. These aspects are incorporated in the table of games via a phenomenological parameter $\alpha \in [0, 1]$, whose lower and higher values are related to bad and good road conditions, respectively. The table of games is described as follows.

● *Interaction with a faster vehicle ($h < k$).* When $h < k$, the candidate vehicle is encountering a faster field vehicle. The result of this interaction can be modeled according to a *follow-the-leader* strategy, which implies that the candidate vehicle either maintains its current speed or possibly accelerates, depending on the available surrounding free space. The dynamics is visualized in Figure 6.5.1a.

$$A_{hk}^i[\rho] = \begin{cases} 1 - \alpha(1 - \rho) & \text{if } i = h, \\ \alpha(1 - \rho) & \text{if } i = h + 1 \quad (h, k = 1, \ldots, n), \\ 0 & \text{otherwise.} \end{cases} \qquad (6.5.7)$$

Note that when $\alpha = 0$ (worst road conditions) the candidate vehicle simply keeps its current speed and does not accelerate in any case; conversely, when $\alpha = 1$ (best road conditions) the result of the interaction is essentially dictated by the local traffic congestion.

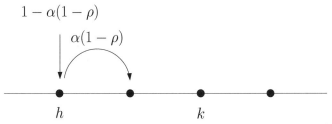

Figure 6.5.1a: Interaction rules of the candidate vehicle with a faster field vehicle.

• **Interaction with a slower vehicle ($h > k$).** When $h > k$, the candidate vehicle interacts with a slower field vehicle. In this case, it is assumed that it does not accelerate and either it is forced to queue, reducing its speed to that of the leading vehicle, or it maintains its current speed, because it has enough free space to overtake, as shown in Figure 6.5.1b.

$$A_{hk}^i[\rho] = \begin{cases} 1 - \alpha(1-\rho) & \text{if } i = k \\ \alpha(1-\rho) & \text{if } i = h \quad (h, k = 1, \dots, n). \\ 0 & \text{otherwise} \end{cases} \qquad (6.5.8)$$

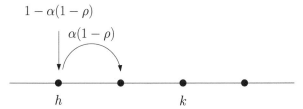

Figure 6.5.1b: Interaction rules of the candidate vehicle with a slower field vehicle.

Note that this choice amounts to defining a *probability of passing $P_\alpha = P_\alpha[\rho]$* dependent on the local traffic and parameterized by the road conditions:

$$P_\alpha[\rho] = \alpha(1-\rho).$$

The emptier the road is, the closer to α this probability becomes, and, if the road conditions allow, the candidate vehicle is more likely to overtake the leading field vehicle without the need to slow down.

• ***Interaction with an equally fast vehicle*** *($h = k$).* When $h = k$, the candidate vehicle and the field vehicle are traveling at the same speed. In this case, the result of the interaction has a higher degree of randomness: the physical situation does not suggest any *a priori* more probable upshot. Therefore, the phenomenological idea of the spread of the velocity is assumed as a guideline: the two vehicles are unlikely to strictly maintain their speed during the motion, for this would imply that they do not interact, behaving as if they were alone on the road. Therefore, the effect of the interaction is distributed over four possible outcomes as shown in Figure 6.5.1c.

$$A_{hh}^i[\rho] = \begin{cases} \alpha\rho & \text{if } i = h - 1 \\ 1 - \alpha & \text{if } i = h \\ \alpha(1 - \rho) & \text{if } i = h + 1 \\ 0 & \text{otherwise} \end{cases} \quad (h = 2, \ldots, n - 1). \qquad (6.5.9)$$

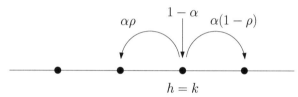

$$h = k$$

Figure 6.5.1c: Interaction rules of the candidate vehicle with a field vehicle of the same velocity.

The parameter α plays the role of a tuning coefficient that regulates the mutual relevance of the outcomes. If $\alpha = 0$, then $A_{hh}^h[\rho] = 1$ for each $h \in \{2, \ldots, n-1\}$ and each ρ, so that one obtains a trivial interaction that does not cause any velocity transition. Conversely, if $\alpha = 1$, the interaction between two equally fast vehicles results in a full spread of the velocity, since $A_{hh}^h[\rho] = 0$ for each ρ, and consequently none of them is allowed to maintain the current speed.

Note that the form of $A_{hh}^i[\rho]$ applies only if $h \neq 1, n$. In contrast, a technical modification is needed at the boundary of the velocity grid, since when $h = 1$ or $h = n$ the candidate vehicle cannot brake or accelerate, respectively, due to the lack of further lower or higher velocity classes. In these cases, the deceleration or the acceleration is merged into the tendency to preserve the current speed:

$$A_{11}^i[\rho] = \begin{cases} 1 - \alpha(1 - \rho) & \text{if } i = 1 \\ \alpha(1 - \rho) & \text{if } i = 2 \\ 0 & \text{otherwise,} \end{cases} \qquad (6.5.10a)$$

and

$$A_{nn}^i[\rho] = \begin{cases} \alpha\rho & \text{if } i = n-1 \\ 1 - \alpha\rho & \text{if } i = n \\ 0 & \text{otherwise.} \end{cases} \qquad (6.5.10b)$$

Concerning the preceding table of games a few comments are in order. Specifically, only velocity classes close to the current one are involved, with the remarkable exception of the interaction with a slower vehicle: in this case, the slowdown leads the candidate vehicle to the velocity v_k of the leading vehicle, no matter how large the difference between h and k.

Moreover, note that the term w weights the interactions of the candidate vehicle with each field vehicle located in front of it in the visibility zone. Different choices of w can be made according to different possible criteria to evaluate the effectiveness of the interactions; however, setting $w(t, y) = \delta_x(y)$ for all $t \geq 0$ allows us to formally recover the localized interactions framework.

The model above can be used to analyze specific traffic flow phenomena. An interesting application is the analysis of the spatially homogeneous problem, in which the distribution function f is independent of the space variable:

$$f = f(t, v) = \sum_{i=1}^n f_i(t)\delta_{v_i}(v), \qquad \frac{\partial f_i}{\partial x} = 0, \quad \forall i = 1, \ldots, n. \qquad (6.5.11)$$

Consequently, the model reduces to

$$\frac{df_i}{dt} = \sum_{h=1}^n \sum_{k=1}^n \eta[\rho] A_{hk}^i[\rho] f_h f_k - f_i \sum_{h=1}^n \eta[\rho] f_h, \qquad i = 1, \ldots, n. \qquad (6.5.12)$$

The mathematical problem consists of the preceding system of ordinary differential equations, in the unknowns $f_i : \mathbb{R}_+ \to \mathbb{R}_+$, with initial conditions

$$f_i(0) = f_i^0 \in \mathbb{R}_+, \quad i = 1, \ldots, n. \qquad (6.5.13)$$

The spatially homogeneous problem is a good benchmark for testing the reliability of the theoretical predictions with respect to the available experimental data, since it provides some information on the trend of the system toward the equilibrium (the **fundamental diagram**) that can be duly compared with the measurements performed under uniform flow conditions which are deeply analyzed by Kerner (2004).

In particular, Figure 6.5.2 shows the behavior of the velocity distribution reached under uniform flow conditions corresponding to the value $\alpha = 1$ of the parameter denoting the quality of the road. Figure 6.5.3 shows the behavior of the flux distribution reached under uniform flow conditions.

Both figures show the transition, well documented by Kerner, that occurs for a critical value of the density. When the quality of the road decreases, namely α attains a lower value, the transition is anticipated at lower values of ρ. This transition, which is experimentally observed, can be used to identify the parameter α related to the quality of the road.

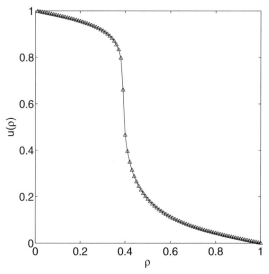

Figure 6.5.2: Average velocity as functions of the macroscopic density for $\alpha = 1$.

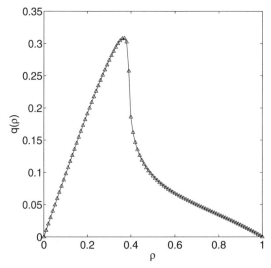

Figure 6.5.3: Macroscopic flux as function of the macroscopic density.

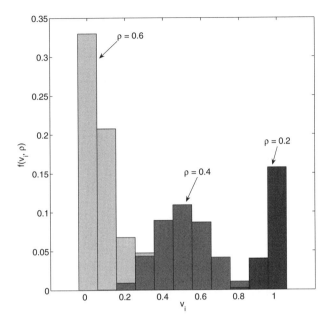

Figure 6.5.4: Equilibrium distribution of the velocity for three possible values of the density ($\rho = 0.2$, $\rho = 0.4$, $\rho = 0.6$) and for $\alpha = 1$.

Figure 6.5.4 provides a further visualization of the simulation showing the velocity distribution for different values of the density. We remark on the concentration of vehicles in the extreme velocity classes for low and high ρ, as opposed to their central distribution for intermediate density.

The paper by Delitala and Tosin (2007) provides proofs of stability of equilibrium solutions for low numbers of the discrete velocities. It is hoped that the proof can be generalized to the case of an arbitrarily large number n.

These encouraging results suggest that we also analyze phenomena in the spatially inhomogeneous problem, which describes the spatial and temporal evolution of the traffic subjected to suitable initial and boundary conditions. In particular, a qualitative analysis is developed addressing some representative cases well documented in the literature, in order to test the ability of the model to reproduce some typical characteristics of traffic flow.

• We consider first the problem of ***formation of a queue***, namely the formation of a queue due to the accumulation of some incoming vehicles behind a pre-existing group of motionless vehicles.

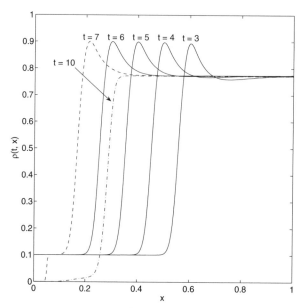

Figure 6.5.5: Evolution of a queue at different times. Dashed line refers to the density profile when vehicles stop entering the domain from the left; dashed-dotted line represents the emptying of the queue as a consequence of the downflow of the vehicles from the right.

As initial conditions all distribution functions f_i, $i \geq 2$, are taken equal to zero, while the one corresponding to the first velocity class $i = 1$ is instead assumed constant in a suitable stretch behind the outflow boundary $x = 1$ of the spatial domain. Hence, f_1 determines for $t = 0$ the initial profile of the density ρ, which exhibits a plateau representing the above-mentioned pre-existing queue. At the inflow boundary $x = 0$ a group of incoming vehicles enters the domain with a certain positive velocity chosen as the maximum possible according to the velocity grid.

Figure 6.5.5 shows the result of the simulation in terms of the macroscopic density ρ. Note in particular the expected enlargement of the plateau due to a backward propagation of the queue toward the inflow boundary, with a nearly constant maximum value of the density located in the rear part of the group of vehicles. A fundamental contribution in correctly reproducing these features is given by both the dislocation of the interactions over the whole amplitude of the visibility zone and the increment in the interaction rate for growing density. Indeed, without the former it is well known that there is no way to obtain backward propagation of the information (see, e.g., Klar and Wegener (2000)), that is, the initial plateau would not grow longitudinally along the road, its rear front remaining fixed at the same initial location for all $t > 0$. On the other hand, without the latter the density may achieve locally values greater than 1; this risk is especially high in the rear part of the queue, at the attack point between the slow queued vehicles and the fast incoming ones, where many incoming vehicles are required to slow down because there is little space left for overtaking. The effect of the interaction rate, which grows to infinity for ρ close to 1, is precisely that of inducing numerous transitions of velocity class by the vehicles, avoiding the superposition of different density waves which may locally sum to more than 1.

At time $t = 7$ the left boundary condition switches to zero for all distribution functions f_i, that is, no vehicle is entering the domain from then on. As a consequence, a slow downflow of the vehicles from the right boundary begins, with a progressive emptying of the queue.

• Let us now consider the ***bottleneck problem*** by analyzing the effect on the traffic of a variation in the maximum density allowed along the road. This situation may arise as a consequence of a reduction in the number of lanes available to the vehicles, or more generally because of a narrowing of the roadway.

A scaling of the nondimensional density ρ according to a variable maximum value, which depends on x as shown by the dotted line in Figure 6.5.6, is introduced. In particular, a bottleneck density profile which is close to 1 at the inflow boundary and decreases to 0.4 at the outflow boundary, causing a reduction of 60% in the road capacity, is visualized.

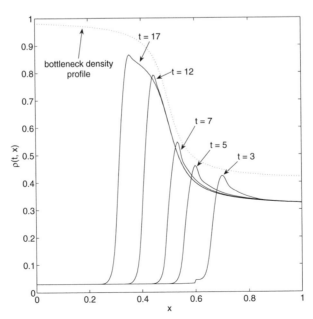

Figure 6.5.6: Formation and evolution of a queue caused by a bottleneck, that is, a variation of the maximum density allowed along the road due

to a reduction in the road capacity. Dotted line represents the bottleneck density profile.

6.6 On the Modeling by Active Particle Methods

The critical surveys reported in the preceding sections have shown that various models based on the methods of mathematical kinetic theory have been developed after the pioneer work by Prigogine. As we have seen, modeling microscopic interactions plays a central role in the derivation of an evolution equation for the one-particle distribution function.

Interactions are modeled by taking into account the drivers' behavior, which modifies the rules of classical mechanics according to the drivers' ability to adjust the dynamics of the vehicles to the real flow conditions. In general, this adjustment is finalized to improve the comfort of the passengers and avoid instabilities which may generate car crashes.

Interactions satisfy conservation of the number of particles, but not momentum and energy; the strategy of drivers dominates the interaction rules, and consequently vehicles may accelerate or decelerate according to such a strategy.

The modeling methods in the literature are not really developed on the basis of methods of kinetic theory for active particles because the explicit role of the activity variable is not considered. However, some preliminary attempts were introduced in the paper by Paveri Fontana (1975), where the interesting concept of a desired velocity program was introduced; and by Delitala (2003), who suggested the use of a random variable to model vehicle interactions.

Namely the mathematical models do not effectively consider Daganzo's remark, reported in Subsection 6.2.4, that modeling should take into account a broad variety of driver-vehicle systems (active particles). According to the conceptual lines offered in this book, a driver-vehicle should be regarded as an active particle, where the activity must be included into the microscopic variable. Moreover, modeling should not be developed according to the continuous kinetic theory, as the number of vehicles is not sufficiently large to apply continuous representation, as critically analyzed in Subsection 6.2.4. See also Shang, Wan, and Kama (2007), who examine the fractal structure of traffic flow configurations.

This section deals with the analysis of some ideas concerning modeling traffic flow phenomena by methods of the kinetic theory for active particles. The analysis developed here does not lead to specific models, but simply to ideas and suggestions to be developed within a proper research program.

The general idea is that the microscopic state of the interacting entities includes, in addition to position and velocity, also the activity variable $u \in [0,1]$, where $u \to 0$ denotes slow vehicles in the hands of nonaggressive drivers, while $u \to 1$ denotes the opposite: fast vehicles in the hands of experienced drivers. Intermediate values may correspond to slow vehicles in the hands of experienced drivers, or fast vehicles in the hands of nonexperienced drivers.

The activity variable plays a role in the modeling of mechanical interactions. This role needs to be specialized for each specific model. For instance, referring to the Delitala and Tosin (DT)-model the role of u can be inserted directly into the table of games through the parameter α, which may be modified by u. The modeling should consider that the quality of the road is made worse, or technically improved, by the technical quality of the driver-vehicle system.

Of course, this oversimplified description should be technically improved with reference to experimental data that also follows the identification procedure of the parameter α given in the preceding section.

The simplest approach to model the role of u on the overall model consists in assuming factorization of the probability density $\varphi(u)$ over u from the distribution function $f^m(x,v)$ over the mechanical variables x,v. However, the probability density φ must be conditioned by the local density

$$f(t,x,v,u) = f^m(t,x,v)\varphi(u;\rho), \qquad (6.6.1)$$

where

$$\int_{D_u} \varphi(u;\rho)\,du = 1, \quad \forall \rho \geq 0. \qquad (6.6.2)$$

Let us now denote by $\varphi_0(u) = \varphi(u;\rho = 0)$ the density corresponding to free flow; empirical data may possibly lead to the identification of the parameter of $\varphi_0(u)$. On the other hand, as ρ increases $\varphi(u;\rho)$ should tend to a Dirac delta over the lowest value of u.

Following some heuristic reasonings, one can model $\varphi(u;\rho)$ by a probability density identified by mean value and variance considered as monotone decreasing functions of ρ, both tending to zero when ρ tends to one. Namely, all driver-vehicles behave in the same way in traffic jam conditions. Finally, the overall model is obtained by using the activity variable u as a random parameter in the table of games and subsequently by linking the above phenomenological model to the evolution equation.

This reasoning can be developed for different types of discrete models, for instance for the CDF and DT models, as well as for some of the continuous models reported in the preceding sections. However, a deeper approach requires modeling microscopic interactions for the whole set of variables identifying the microscopic state, for instance again using ρ as a key variable which amplifies or reduces the role of the activity u. As already mentioned, these ideas only provide some preliminary hints for specific research projects still waiting to be properly developed.

7

Complex Biological Systems: Mutations and Immune Competition

7.1 Introduction

One of the most interesting and challenging fields for applying the mathematical kinetic theory for active particles is the modeling of biological systems; specifically, the competition between cell carriers of a pathology and immune cells. See Dunn, Bruce, Ikeda, Old, and Schreiber (2002). These systems substantially differ from those in Chapters 5 and 6 due to a significant presence of proliferative and/or destructive events. This applies to several systems in biology, for instance multicellular systems in which the active particles may be carriers of a pathology. When the pathology refers to cancer cells, which due to genetic mutations have lost their programmed death ability, then one talks about *progressing cells*.

The pathology in this case is the output of various genetic mutations which generate new cells with an increasing degree of malignancy. After various genetic mutations, cancer cells have the ability of expressing biological functions which can allow them to escape from the immune system despite the sentinel guard which should, in principle, contrast progressing cells, see Lollini, Motta, and Pappalardo (2006), Vogelstein and Kinzler (2004), Blankenstein (2005).

A biological function is not generally the same for all cells, as it is statistically distributed. This biological event generates what is called *heterogeneity* related to *progression* and to *immune activation* as documented by the papers by Greller, Tobin, and Poste (1996), and by Nowell (2002). Models should have the ability to describe the progression of cancer cells and their competition from immune cells which express, unless inhibited, their antagonistic ability.

The modeling analyzed in this chapter refers to a system homogeneous in space, although suggestions for dealing with space problems are given. The technical differences with respect to the preceding chapter are as follows.

i) The number of active particles evolves in time due to proliferation and destruction phenomena.

ii) The number of governing equations grows in time due to the onset of new populations generated by genetic mutations.

It is well known that lack of mass conservation and a variable number of equations generate various technical difficulties in the qualitative and computational analysis of solutions of mathematical problems as documented in Chapter 4 of the book by Bellouquid and Delitala (2006).

Application of methods of mathematical kinetic theory to model the immune competition with special attention to cancer phenomena was initiated by Bellomo and Forni (1994) and subsequently developed by various authors. Recent contributions are due to Derbel (2004), De Angelis and Jabin (2003), (2005), Kolev (2005), Kolev, Kozlowska, and Lachowicz (2005), and Bellouquid and Delitala (2004), (2005) among others. Several interesting results have been organized in the book by Bellouquid and Delitala (2006), which discusses several aspects of this competition.

This literature refers to biological situations where cancer cells are already present in the environment. In other words, models do not include the description of genetic mutations which generate progressing cells with a greater degree of malignancy. This biological phenomenon has been studied in a recent paper by Delitala and Forni (2007), which offers a useful background to this chapter.

The chapter is organized through four more sections as follows.

– Section 7.2 provides a phenomenological description of the biological system. From the viewpoint of biological sciences, it defines the statistical representation, which is assumed to be continuous, over the microscopic state of interacting cells.

– Section 7.3 deals with the selection and a critical analysis of the mathematical structures used for modeling the immune competition between progressing and immune cells in both cases: when progressing cells do not have the ability to generate new particles in a third population, and when this ability is expressed. As we shall see, modeling this ability provides a richer description of biological reality with respect to the first case.

– Section 7.4 analyzes a particular model showing its ability to describe some specific aspects of the immune competition. Specifically, the model shows how both outputs can be described according to a proper selection of the parameters: weakening and destruction of tumor cells due to an active immune system, or progressive inhibition of the immune system and blow up of tumor cells.

– Section 7.5 offers a critical analysis to indicate research perspectives essentially concerning further modeling issues and, in particular, the possibility of introducing the space variable in the microscopic state. This development opens a window for studying the difficult problem of deriving macroscopic equations suitable to describe the evolution of tissues within the framework of continuum mechanics.

7.2 Modeling the Immune Competition

This section gives a phenomenological description of multicellular systems and of immune competition which may be useful in developing the models reported in the following sections.

Researchers in the field of biological sciences should be patient with the naive description delivered here. It suffers because 1) the author, an applied mathematician, lacks an in-depth knowledge of biological sciences, and 2) it has required a constant effort to produce the mathematical equations. This effort was necessary to constrain a complex physical reality into structured frameworks, although this process is so difficult as to appear impossible. The reader can find in the excellent book by Weinberg (2006) all necessary information on the biology of the system under consideration.

Referring to the paper by Hartwell, Hopfield, Leibner, and Murray (1999) as a relevant background for our reasoning here, the modeling process must take into account the peculiarity of living systems which do not obey the laws of physics and chemistry in the same way as inert matter. The notion of function or purpose differentiates biology from other natural sciences. What really distinguishes biology from physics is the ability of biological systems to survive and reproduce in connection with the expression of specific functions.

Biological functions can modify the conservation laws of classical mechanics due to the organized behavior of the several entities composing a living system. It follows that the mathematical kinetic theory for active particles is the natural candidate to offer mathematical tools for modeling complex biological systems.

Bearing all this in mind, consider a biological system constituted by several cell populations interacting *in vivo*. The biological phenomena to be modeled are the onset of cancer cells and their competition (or simply interactions) with the immune system. The characterization of the system suggests the identification of three natural scales: processes on the ***cellular scale*** are triggered by signals stemming from the ***subcellular level***, say the ***molecular scale***, and have an impact on the ***macroscopic scale***, namely on the organism, as tumor cells condense into solid forms which grow and are spread into the organism. The modeling developed in this chapter refers to the cellular scale; however, mathematical models developed at this scale should retain suitable information from the lower cellular (molecular) scale, and they should allow the derivation of macroscopic equations by suitable asymptotic limits related to condensation and fragmentation events.

A brief description of the relevant biological phenomena at the first two scales is useful.

The molecular scale: The evolution of a cell is regulated by the genes contained in its nucleus. Receptors on the cell surface receive signals which are then transduced to the cell nucleus, where various genes are activated or suppressed. Particular signals can induce a cell to reproduce itself in the form of identical descendants (called clonal expansion), or to die and disappear apparently without trace (apoptosis or programmed cell death). Clonal expansion of tumor cells activates a competitive-cooperative interaction between them and the cells of the immune system. If the immune system is active and able to recognize the tumor cells, then it may be able to destroy the incipient tumor; otherwise, tumor growth develops progressively. The activation and deactivation of immune cells, too, is regulated by cytokine signals delivered by cells of the immune system or by the tumor, see Blankenstein (2005).

The cellular scale: Models are developed to simulate the effects of the failure of programmed cell death and of the loss of cell differentiation. If and when a tumor cell is recognized by immune cells, a competition starts which results either in the destruction of tumor cells or in the inhibition and depression of the immune system. Cellular interactions are regulated by signals emitted and perceived by cells through complex reception and transduction processes, see Blume-Jensen and Hunter (2001). Therefore, the connection to the subcellular scale is evident. However, the development of tumor cells, if not suppressed by the immune system, tends towards condensation into a solid form so that macroscopic features become important.

This chapter refers to modeling on the above scales, although it has to be kept in mind that after a suitable maturation time, tumor cells may start to condense and aggregate into an entity which eventually evolves as a *quasi-fractal surface* which interacts with the outer environment, for example, normal host cells and the immune system. These interactions usually occur on the surface and within a layer where angiogenesis (the process of formation of new blood vessels, induced by factors secreted by the tumor and vital for tumor growth) takes place, see Folkman (2002).

To reduce complexity, the modeling developed in the following sections refers to the theory of modules proposed by Hartwell. A module is an assembly of biological elements (or even systems) which has the ability to develop a well-defined biological function. The class of models dealt with in this chapter identifies a module as a cell population which, collectively and cooperatively, has the ability to develop a certain function, which differs from population to population.

It is well understood that the immune system consists of several cell populations, e.g., lymphocytes, leucocytes, and macrophages, which have different abilities from the identification of progressing cells to their mechanical destruction. Similarly progressing cells may be viewed as belonging to different populations related to different levels of genetic mutations.

7.3 Mathematical Structures for Modeling

This section discusses the selection of the mathematical frameworks to be used for the modeling of the system described in Section 7.2.

The modeling is developed at the cellular scale using the methods of mathematical kinetic theory for active particles. The overall state of the system is described by the statistical distribution, for each cell population, over the state of cells corresponding to biological functions. Macroscopic quantities are obtained by suitable averaging processes.

We derive the mathematical equations to describe the evolution of the distribution function by using conservation equations in the elementary volume of the space of microscopic states. The rate of increase of cells in this volume, i.e., of the number of cells with a certain biological state, is the output of cell interactions which modify the mechanical and biological state, and includes reproduction and destruction events, as well as the onset of new progressing cells with a higher degree of malignancy.

This reasoning can be immediately transferred into suitable mathematical terms according to Chapter 2. Let us then consider a system constituted by p interacting cell populations labeled by the indexes $i = 1, \ldots, p$. The physical *activity* variable used to describe the state of each cell is denoted by u, which refers specifically to the main *biological function* expressed by the cell population. The description of the overall state of the system is defined by the one-cell generalized distribution function

$$f_i = f_i(t, u), \quad [0, T] \times D_u \to \mathbb{R}_+ , \qquad (7.3.1)$$

for $i = 1, \ldots, p$, and such that $f_i(t, u)\, du$ denotes the number of cells whose state, at time t, is in the interval $[u, u+du]$, while the domain of the activity variable is denoted by D_u.

If f_i is known, then macroscopic gross variables can be computed, under suitable integrability properties, as moments weighted by the above distribution function. Focusing on the macroscopic variables of interest for the class of biological systems under consideration, the **number density**, also called the **size**, of the ith population is given by

$$n_i[f_i](t) = \int_{D_u} f_i(t, u) \, du \,, \tag{7.3.2}$$

while the total number density of all populations is given by the sum of all n_i

$$n[f](t) = \sum_{i=1}^{p} n_i(t) \,. \tag{7.3.3}$$

Note that the number of cells can now locally grow or decay in time, and growth may also result in blow-up behavior.

Focusing more specifically on biological functions, linear moments related to the ith populations can be called the **activation** at time t, and are computed as follows:

$$A_i = A[f_i](t) = \int_{D_u} u \, f_i(t, u) \, du \,, \tag{7.3.4}$$

while the corresponding **activation density** is given by

$$\mathcal{A}_i = A[f_i](t) = \frac{A[f_i](t)}{n_i(t)} = \frac{1}{n_i[f_i](t)} \int_{D_u} u \, f_i(t, u) \, du \,. \tag{7.3.5}$$

This action can also be expressed by quadratic quantities analogous to energy:

$$E_i = A[f_i](t) = \int_{D_u} u^2 \, f_i(t, u) \, du \,, \tag{7.3.6}$$

while the corresponding **quadratic activation density** is given by

$$\mathcal{E}_i = A[f_i](t) = \frac{A[f_i](t)}{n_i(t)} = \frac{1}{n_i[f_i](t)} \int_{D_u} u^2 \, f_i(t, u) \, du \,. \tag{7.3.7}$$

Both the linear and quadratic quantities have a relevant biological meaning as the activities define the overall action expressed by the cells of a certain population. This action may be the output of a weak action of single cells, but it may possibly be applied by a large number of cells. However, the density defines more precisely the action expressed, in the mean, by each cell. The specific meaning differs for each population: progression ability for cancer cells, defense ability for immune cells.

As we have seen in Chapter 2, interactions are of short or long range type. The analysis developed here refers to short range interactions occurring between the **test**, or **candidate**, and the **field** cell. Encounters between pairs of cells can be classified as **conservative interactions**, which modify the state, mechanical and/or biological, of the interacting cells, but not the size of the population; and **proliferative** or **destructive interactions**, which generate death or birth of cells due to pair interactions. The first framework is limited to the case where proliferative interactions do not generate a cell in a population different from the interacting pairs.

The dynamics of the interactions has been already visualized in Chapter 2, with reference to the figures of Section 2.3, and some additional representation is given in the next section referring to a specific model. The mathematical selected structure corresponding to spatial homogeneity is as follows:

$$\frac{\partial f_i}{\partial t}(t, u) = J_i[f](t, u) = C_i[f](t, u) + D_i[f](t, u).\qquad(7.3.8)$$

The domain D_u of the microscopic state, in view of the following applications, is assumed to coincide with the positive real axis: $u \in [-\infty, \infty)$. Moreover:

• $C_i[f](t, u)$ models the flow, at time t, into the elementary volume of the state space of the ith population due to conservative interactions. Technical calculations, reported in Chapter 2, yield

$$C_i[f](t, u) = \sum_{j=1}^{p} \eta_{ij} \int_{-\infty}^{\infty} \int_{-\infty}^{\infty} \mathcal{B}_{ij}(u_*, u^*; u) f_i(t, u_*) f_j(t, u^*) \, du_* \, du^*$$

$$- f_i(t, u) \sum_{j=1}^{p} \eta_{ij} \int_{-\infty}^{\infty} f_j(t, u^*) \, du^*,\qquad(7.3.9)$$

where η_{ij} is the encounter rate, referred to encounters of **candidate particles**, with state u_* in the ith population and **field particles**, with state u^* in the jth population. $\mathcal{B}_{ij}(u_*, u^*; u)$ denotes the probability density that the candidate particles will fall into the state u remaining in the same populations. Conservative equations modify the microscopic state, but not the number of cells.

• $D_i[f](t, u)$ models the net flow, at time t, into the elementary volume of the state space of the ith population due to proliferative and destructive interactions without a transition of population:

$$D_i[f](t, u) = f_i(t, u) \sum_{j=1}^{p} \eta_{ij} \int_{-\infty}^{\infty} \mu_{ij}(u, u^*) f_j(t, u^*) \, du^* , \qquad (7.3.10)$$

where $\mu_{ij}(u, u^*)$ models the net flux within the same population due to interactions, which occur with rate η_{ij}, of the **test particle**, with state u, of the ith population and the **field particle**, with state u^*, of the jth population.

Substituting the above expression into (7.3.8) yields

$$\frac{\partial f_i}{\partial t}(t, u) = \sum_{j=1}^{p} \eta_{ij} \int_{-\infty}^{\infty} \int_{-\infty}^{\infty} \mathcal{B}_{ij}(u_*, u^*; u) f_i(t, u_*) f_j(t, u^*) \, du_* \, du^*$$

$$- f_i(t, u) \sum_{j=1}^{p} \eta_{ij} \int_{-\infty}^{\infty} f_j(t, u^*) \, du^*$$

$$+ f_i(t, u) \sum_{j=1}^{p} \eta_{ij} \int_{-\infty}^{\infty} \mu_{ij}(u, u^*) f_j(t, u^*) \, du^* . \qquad (7.3.11)$$

Let us now consider the case of cells which have the ability of generating new cells in a different population. Then the mathematical structure (7.3.8) must be modified as follows:

$$\frac{\partial f_i}{\partial t}(t, u) = Q_i[f](t, u) = J_i[f](t, u) + P_i[f](t, u) , \qquad (7.3.12)$$

where $P_i[f](t, u)$ models the flow, at time t, into the elementary volume of the state space of the ith population due to proliferative interactions with a transition of population:

$$P_i[f](t, u) = \sum_{h=1}^{p} \sum_{k=1}^{p} \eta_{hk} \int_{-\infty}^{\infty} \int_{-\infty}^{\infty} \mu_{hk}^i(u_*, u^*; u) f_h(t, u_*) f_k(t, u^*) \, du_* \, du^* ,$$

$$(7.3.13)$$

where $\mu_{hk}^i(u_*, u^*; u)$ models the net proliferation into the ith population due to interactions, which occur with rate η_{hk}, of the **candidate particle**, with state u_*, of the hth population and the **field particle**, with state u^*, of the kth population.

The corresponding general structure is as follows:

$$\frac{\partial f_i}{\partial t}(t,u) = \sum_{j=1}^{p} \eta_{ij} \int_{-\infty}^{\infty} \int_{-\infty}^{\infty} \mathcal{B}_{ij}(u_*,u^*;u)f_i(t,u_*)f_j(t,u^*)\,du_*\,du^*$$

$$- f_i(t,u) \sum_{j=1}^{p} \eta_{ij} \int_{-\infty}^{\infty} f_j(t,u^*)\,du^*$$

$$+ \sum_{h=1}^{p} \sum_{k=1}^{p} \eta_{hk} \int_{-\infty}^{\infty} \int_{-\infty}^{\infty} \mu_{hk}^{i}(u_*,u^*;u)f_h(t,u_*)f_k(t,u^*)\,du_*\,du^*$$

$$+ f_i(t,u) \sum_{j=1}^{p} \eta_{ij} \int_{-\infty}^{\infty} \mu_{ij}(u,u^*)f_j(t,u^*)\,du^*, \tag{7.3.14}$$

where now the number of populations m is greater than p.

These structures act as a paradigm for the derivation of specific models to be obtained by a detailed modeling of microscopic interactions generating well-defined expressions of the terms η, $\mathcal{B}$, and μ. Moreover, to reduce complexity, each population is identified by the ability of expressing one function only; hence, the number of populations is a consequence. According to this assumption, each population acts as a module which expresses their biological function within the framework of the above stochastic representation.

Figures 7.3.1 and 7.3.2 visualize genetic mutations which generate progressing cells characterized by uncontrolled mitosis.

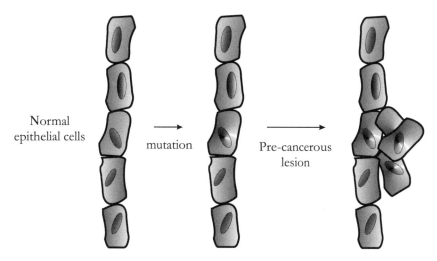

Normal epithelial cells → mutation → Pre-cancerous lesion

Fig. 7.3.1: Genetic mutations from normal to pre-cancerous cells.

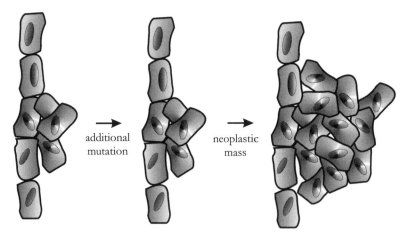

Fig. 7.3.2: Additional mutations producing neoplastic cells.

7.4 An Example of Mathematical Models

This section shows how the mathematical frameworks of Section 7.3 can be properly particularized to generate models suitable to describe the onset of cancer cells and their evolution in competition with the immune system. To develop the analysis of the highly complex system under consideration requires much more space than a single chapter of this book. Therefore, only an example of a model is presented and critically analyzed to stimulate research activity in the field. Specific research perspectives are analyzed in Section 7.5.

The activity variable which describes the biological function of cancer cells is called **progression**, and is identified by a scalar variable with positive values denoting how far cells are from the normal state. Increasing values of u mean moving towards greater values of the proliferation ability up to tissue-invasive and metastatically competent states. Immune cells are characterized by a different biological function, which corresponds to their ability to recognize and distinguish progressing cells, possibly destroying them due to a cooperative action of all populations of the immune system. The variable u is also defined over negative values corresponding to biological functions opposite to the ones indicated above.

For both cell populations, a relevant biological characteristic, well remarked in the paper by Greller, Tobin, and Poste (1996), is the **heterogeneity** over the activity variable, which means that biological functions are not the same for all cells, but are statistically distributed (for progression) as shown in Figure 7.4.1. Heterogeneity also occurs in space as cells develop their progression state locally so that the statistical representation differs from place to place. An analogous representation corresponds to the biological function of immune cells.

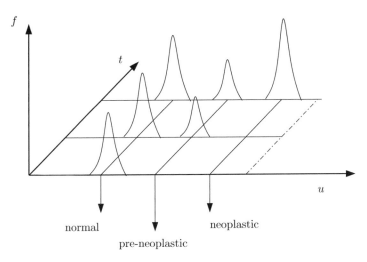

Fig. 7.4.1: Time evolution of progression and heterogeneity.

The existing literature is documented in the book by Bellouquid and Delitala (2006). Various models proposed in the book describe the evolution in time of the distribution functions over the activity variables of all interacting cell populations. The competition may end either with the blow up of cancer cells after at least partial inhibition of immune cells; or with the destruction of cancer cells due to the action of the immune system which remains active.

A simple model is proposed here, referring to the mathematical structure (7.3.10), as a technical interpretation of the above-cited book. Then some developments are analyzed. Our simple model is described mainly with tutorial aims. The reader interested in developing a research program in the field can improve the model by increasing its descriptive ability, taking advantage of several indications delivered by Bellouquid and Delitala (2006), as well as of some interesting developments recently proposed by Delitala and Forni (2007), where a model which describes sequential genetic mutations with increasing level of malignancy is developed.

Let us consider a system constituted by two populations whose microscopic state $u \in [-\infty, \infty)$ has a different meaning for each population:

$i = 1$: ***Environmental cells*** - The state u refers to the **natural** state (normal endothelial cells) for negative values of u; to the abnormal state, i.e., cells which have lost their differentiated state and become progressing cells, for positive values of u, with the additional ability to inhibit immune cells.

$i = 2$: **Immune cells** - Negative values of u correspond to nonactivity or *inhibition*; positive values of u to **activation** and hence their ability to **contrast** the growth of tumor cells;

Before providing a detailed description of the microscopic interactions, some preliminary assumptions which reduce the complexity of the system described by Eq. (7.3.13) are stated:

H.1. The number and distribution of cells of the first population is denoted by $f_1(0, u) = f_{10}(u)$, and the distribution function of all cell populations are normalized with respect to the initial number density of normal cells of the first population:

$$n_{10}^E = \int_{-\infty}^{0} f_{10}(u)\, du\,. \tag{7.4.1}$$

H.2. The distribution over the velocity variable is constant in time, therefore, the encounter rate is constant for all interacting pairs. For simplicity it is assumed: $\eta_{ij} = 1$, for all i, j.

H.3. The term $\mathcal{B}_{ij}$, related to the transition probability density, is assumed to be defined by a given delta distribution identified by the most probable output $m_{ij}(u_*, u^*)$, which depends on the microscopic state of the interacting pairs:

$$\mathcal{B}_{ij}(u_*, u^*; u) = \delta(u - m_{ij}(u_*, u^*))\,. \tag{7.4.2}$$

Moreover, it is useful to introduce, for technical calculations, the stepwise function, $U_{[a,b]}(z)$ such that $U_{[a,b]}(z) = 1$, if $z \in [a, b]$ and $U_{[a,b]}(z) = 0$, if $z \notin [a, b]$.

Following these preliminary assumptions, a detailed description of the nontrivial interactions, i.e., those assumed to play a role in the evolution of the system, can now be given according to phenomenological models. Interactions which do not affect the evolution of the system are not reported.

• **Conservative interactions**

C.1. Interactions between cells of the first population generate a continuous trend in this population towards progressing states identified by the most probable output:

$$m_{11} = u_* + \alpha_{11}\,,$$

where α_{11} is a parameter related to the inner tendency of both a normal and abnormal cell to degenerate.

C.2. The most probable output of the interaction between an active immune cell with a progressing cell is given as follows:

$$u_* \geq 0\,, u^* \geq 0 : \quad m_{21} = u_* - \alpha_{21}\,,$$

where α_{21} is a parameter which indicates the ability of abnormal cells to inhibit immune cells.

● **Proliferative-destructive interactions**

PD.1. Progressing cells undergo uncontrolled mitosis stimulated by encounters with nonprogressing cells due to their feeding ability:

$$\mu_{11}(u, u^*) = \beta_{11} U_{[0,\infty)}(u) U_{(-\infty,0)}(u^*),$$

where β_{11} is a parameter which characterizes the proliferative ability of abnormal cells.

PD.2. Active immune cells proliferate due to encounters with progressing cells:

$$\mu_{21}(u, u^*) = \beta_{21} U_{[0,\infty)}(u) U_{[0,\infty)}(u^*), \qquad (7.4.3)$$

where β_{21} is a parameter which characterizes the proliferative ability of immune cells.

PD.3. Progressing cells are partially destroyed due to encounters with active immune cells:

$$\mu_{12}(u, u^*) = -\beta_{12} U_{[0,\infty)}(u) U_{[0,\infty)}(u^*),$$

where β_{12} is a parameter which characterizes the destructive ability of active immune cells.

Based on the above modeling of cell interactions, the evolution equation Eq. (7.3.14) generates the following model:

$$\begin{cases} \dfrac{\partial f_1}{\partial t}(t, u) = [f_1(t, u - \alpha_{11}) - f_1(t, u)] \displaystyle\int_{-\infty}^{\infty} f_1(t, u)\, du \\[2mm] \quad + f_1(t, u)\beta_{11} \displaystyle\int_{-\infty}^{0} f_1(t, u)\, du - (1 + \beta_{12}) \displaystyle\int_{0}^{\infty} f_2(t, u)\, du\, U_{[0,\infty)}(u), \\[4mm] \dfrac{\partial f_2}{\partial t}(t, u) = f_2(t, u + \alpha_{21}) \displaystyle\int_{0}^{\infty} f_1(t, u)\, du\, U_{[0,\infty)}(u + \alpha_{21}) \\[2mm] \quad + (\beta_{21} - 1) f_2(t, u) \displaystyle\int_{0}^{\infty} f_1(t, u)\, du\, U_{[0,\infty)}(u). \end{cases}$$

$$(7.4.4)$$

The corresponding densities are computed as follows.

$$n_1(t) = \int_{-\infty}^{\infty} f_1(t, u)\, du, \qquad n_2(t) = \int_{-\infty}^{\infty} f_2(t, u)\, du, \qquad (7.4.5)$$

are the zeroth-order moments representing the densities of each cell population;

$$n_1^E(t) = \int_{-\infty}^{0} f_1(t, u)\, du\,, \quad n_1^T(t) = \int_{0}^{\infty} f_1(t, u)\, du\,, \tag{7.4.6}$$

are the densities of normal endothelial and abnormal cells; and

$$n_2^I(t) = \int_{-\infty}^{0} f_2(t, u)\, du\,, \quad n_2^A(t) = \int_{0}^{\infty} f_2(t, u)\, du\,, \tag{7.4.7}$$

are the densities of inhibited and active immune cells.

The model is characterized by five positive phenomenological parameters, small with respect to one:

α_{11} refers to the tendency of endothelial cells to degenerate,

α_{21} refers to the ability of abnormal cells to inhibit the active immune cells,

β_{11} refers to the proliferation rate of abnormal cells,

β_{12} refers to the ability of immune cells to destroy abnormal cells,

β_{21} refers to the proliferation rate of immune cells.

The α-parameters are related to conservative encounters, the β-parameters to proliferative and destructive interactions.

It is useful to show that the evolution equations for the densities cannot be given in a closed form. Specifically, the evolution for the densities $n_1^T(t)$ and $n_1^E(t)$ are obtained by integration of the first equation of (7.4.4) respectively on $\mathbb{R}^+$ and on $\mathbb{R}^-$:

$$\begin{cases} \dfrac{\partial n_1^T(t)}{\partial t} = n_1(t) \displaystyle\int_{-\alpha_{11}}^{0} f_1(t, u)\, du \\[2mm] \qquad\qquad + n_1^T(t)\left[\beta_{11} n_1^E(t) - \beta_{12} n_2^A(t)\right]\,, \\[3mm] \dfrac{\partial n_1^E(t)}{\partial t} = -n_1(t) \displaystyle\int_{-\alpha_{11}}^{0} f_1(t, u)\, du\,. \end{cases} \tag{7.4.8}$$

Applying the same procedure to the second equation of (7.4.4) yields the evolution equation for $n_2^A(t)$ and $n_2^I(t)$:

$$\begin{cases} \dfrac{\partial n_2^A(t)}{\partial t} = n_1^T(t)\left[\beta_{21} n_2^A(t) - \displaystyle\int_{0}^{\alpha_{21}} f_2(t, u)du\right]\,, \\[3mm] \dfrac{\partial n_2^I(t)}{\partial t} = n_1^T(t) \displaystyle\int_{0}^{\alpha_{21}} f_2(t, u)\, du\,. \end{cases} \tag{7.4.9}$$

It is useful to analyze the influence of the parameters of the model and of the mathematical problem over the following two different behaviors:

i) Blow up of progressing cells which are not sufficiently contrasted by immune cells due both to the fast progression of tumor cells and to the weak proliferation of immune cells,

ii) Destruction of progressing cells due to the action of the immune system which has a sufficient proliferation rate.

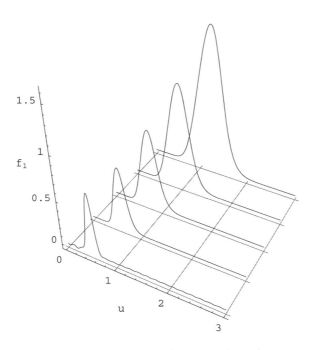

Fig. 7.4.2: Heterogeneity and progression of tumor cells.

Simulations can be focused to visualize the role of the parameters over the output of the competition, for instance, to show how different progression rates can lead to different outputs of the competition.

The two different behaviors are visualized in Figures 7.4.2–7.4.7 according to a different selection of the parameters. Figure 7.4.2 shows how the distribution function of tumor cells moves towards greater values of progression. Correspondingly, immune cells are progressively inhibited as shown in Figure 7.4.3. The evolution of the densities is reported in Figure 7.4.4 which shows the blow up of tumor cells (continuous line) and the progressive inactivation of immune cells (dotted line). The opposite behavior is shown in Figures 7.4.5–7.4.

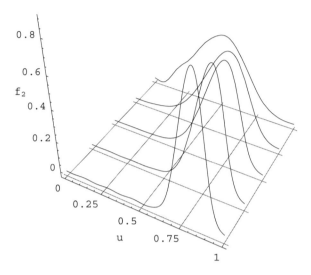

Fig. 7.4.3: Progressive inhibition of immune cells.

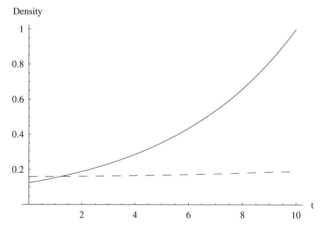

Fig. 7.4.4: Blow up of progressing cells and inactivation of immune cells.

The mathematical model reported above already shows the ability to describe some interesting biological phenomena and specifically the time evolution of heterogeneity and the role of a few biological parameters on

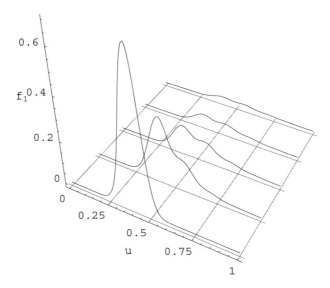

Fig. 7.4.5: Heterogeneity and progressive destruction of tumor cells.

the output of the competition. But it is not able to describe phenomena such as successive genetic selective mutations which generate new cells with relatively more aggressive behavior. This issue is discussed in the final section.

7.5 Modeling Developments and Perspectives

The mathematical model reported in Section 7.4 is a simple application developed by considering only a small part of the large variety of phenomena related to the complex dynamics of the immune competition. The model has been proposed, with tutorial aims, to show an immediate application of the modeling approach. This section provides a critical analysis focused on conceivable research perspectives. Specifically, we introduce some generalizations to enlarge the number of phenomena to be described by mathematical equations.

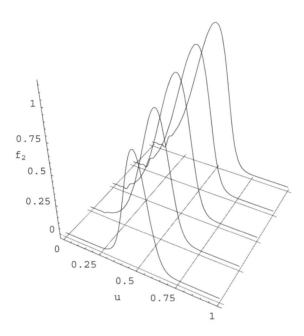

Fig. 7.4.6: Immune cells remain active.

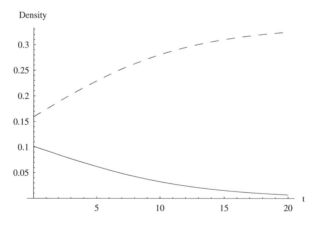

Fig. 7.4.7: Progressive decay of the number density of tumor cells.

Of course, increasing the range of predicted events also increases the complexity of models. Therefore, the model may lose its predictive ability if there are too many parameters to be identified by empirical data.

Therefore, the following issues, among several conceivable ones, are analyzed:

i) Increasing the number of populations to include the description of additional physical phenomena;

ii) Parameter identification possibly developing a mathematical theory of complex biological systems;

iii) Additional modeling of heterogeneity phenomena including the space variable in the microscopic state;

The analysis developed in the following paragraphs is concise; also, some of these topics will be further considered in the last chapter.

• *Models with an increasing number of populations*

The mathematical model described in Section 7.4 has been derived within the framework defined by Eq. (7.3.11) which does not include the modeling of genetic mutations, Hanahan and Weinberg (2000), Volgestein and Kinzler (2004). It follows that the model can describe biological phenomena along a short time interval, whereas for longer times genetic mutations can play a relevant role in the evolution of the system. Moreover, the model has been derived for a small number of populations (regarded as modules expressing a well-defined biological function), so that their mathematical structures do not appear too complex for a computational analysis.

This modeling approach must definitely be regarded as an approximation of physical reality considering that biological functions are the output of the collective behavior of several different types of cells. A relatively more refined interpretation of the theory can identify the function expressed by each type of cell, so that the number of populations is consistently increased. The generalization can be technically developed by referring to the fact that the onset of new populations is related to genetic mutations, which, according to well-accepted theories, are of a finite number. A similar reasoning can be developed for the modeling for immune cells; one looks at the immune system for each specific population rather than modeling them as one module only

The down side of this generalization is that the number of parameters also increases and that their identification may become very difficult or even impossible. For instance, the identification process described below can be practically applied only for models with a number of populations comparable with the one for the model of Section 7.4.

The number of populations can be increased, according to Delitala and Forni (2007) to include new populations with increasing degrees of malignancy related to subsequent genetic transitions. (For instance, with reference to Figures 7.3.1 and 7.3.2, the increasing degrees correspond to mutated dormant cells, to pre-cancerous cells, and to neoplastic cells. The modeling of the onset of cells into new populations can be developed as we have seen in the second model. Of course, additional parameters are needed.

Like the previous one, this model can also be referred to Hartwell's theory of modules, as it corresponds to assuming that genetic modifications are identified at different stages by discrete variables and that each population is regarded as a module which expresses the same type of biological function.

Simulations should be developed to show how this model has the ability to describe the onset of progressing cells and their heterogeneity over the activity variable. Progression corresponds to a sequence of genetic mutations, while heterogeneity corresponds to the presence of different types of mutations. Both phenomena are contrasted by the immune system.

• *Parameter identification problems*

The mathematical models reported in Section 7.4 are characterized by a small number of phenomenological parameters which can possibly be identified by suitable comparisons with experimental data. It is not a simple task not only because of the technical difficulties of the experiments, but because of the different behaviors of the immune competition *in vivo*, see Gillet (2005). Therefore, a competition involving cells with the same biological characteristics can, *in vivo*, generate different behaviors.

As we have seen, the above models attempt to deal with randomness by assuming that the output of the interactions is not deterministic and occurs with a variance.

The process of identification of the parameters can be developed by isolating a small number of biological phenomena with respect to the whole variety delivered by the model. For instance, models can be used to analyze the competition in the case of a suppressed immune system. Therefore, the parameters related to the immune competition, when the immune system is active, can be identified once the other ones have been fixed.

Although the model has the ability to describe interesting biological phenomena and its parameters can possibly be identified, we want to know what additional work must be done to obtain a bio-mathematical theory. The crucial passage is the modeling of the terms η, φ, $\mathcal{B}$, and μ, which describe microscopic interactions as functions of the cell states.

It has been shown how these terms can be modeled according to a phenomenological interpretation of a certain biological system. However, only when this type of information can be delivered by a proper theory consistent with biological sciences will the link between mathematical and biological sciences be complete and a bio-mathematical theory be generated.

• *Modeling space heterogeneity phenomena*

Consider finally the problem of modeling space phenomena considering that heterogeneity occurs with different distributions also depending on the space variable. A simple way to deal with this problem is to assume that the motion is described by random walk models as reported by Eq. (3.4.6) of Chapter 3. This approach has been used in the paper by Bellomo and Bellouquid (2006) to derive macroscopic equations from an underlying microscopic description.

The generalization to the description of space phenomena is written as follows:

$$
\frac{\partial f_i}{\partial t}(t, \mathbf{x}, \mathbf{v}, u) + \mathbf{v} \cdot \nabla_{\mathbf{x}} f_i(t, \mathbf{x}, \mathbf{v}, u)
$$

$$
= \sum_{j=1}^{p} \eta_{ij} \int_{D_{\mathbf{v}}} \left[T_{ij}(\mathbf{v}, \mathbf{v}^*) f_j(t, \mathbf{x}, \mathbf{v}^*, u) - T_{ji}(\mathbf{v}^*, \mathbf{v}) f_i(t, \mathbf{x}, \mathbf{v}, u) \right] d\mathbf{v}^*
$$

$$
+ \sum_{j=1}^{p} \eta_{ij} \int_{0}^{\infty} \int_{0}^{\infty} \mathcal{B}_{ij}(u_*, u^*, u) f_i(t, \mathbf{x}, \mathbf{v}, u_*) f_j(t, \mathbf{x}, \mathbf{v}, u^*) du_* du^*
$$

$$
- f_i(t, \mathbf{x}, \mathbf{v}, u) \sum_{j=1}^{p} \eta_{ij} \int_{0}^{\infty} f_j(t, \mathbf{x}, \mathbf{v}, u^*) du^*
$$

$$
+ \sum_{h=1}^{p} \sum_{k=1}^{p} \eta_{hk} \int_{0}^{\infty} \int_{0}^{\infty} \mu_{hk}^{i}(u_*, u^*; u) f_h(t, \mathbf{x}, \mathbf{v}, u_*) f_k(t, \mathbf{x}, \mathbf{v}, u^*) du_* du^*
$$

$$
+ f_i(t, \mathbf{x}, \mathbf{v}, u) \sum_{j=1}^{p} \eta_{ij} \int_{0}^{\infty} \mu_{ij}(u, u^*) f_j(t, \mathbf{x}, \mathbf{v}, u^*) du^*. \tag{7.5.1}
$$

The above structure, with notations already defined in Section 3.4 of Chapter 3, corresponds to the assumptions that active particles choose any direction with bounded velocity. The structure can be particularize α for instance, according to one of the models described in Section 7.3.

The model described by the above equation is valid only for small perturbations of steady state with the velocity of cells close to zero. It can be used to derive macroscopic equations from an underlying microscopic description. Further developments should consider the possibility, which has a biological meaning, that the turning operator also depends on the activity variable which selects the direction of the movement according to specific strategies. However, this development applied to models derived according to the framework (7.5.1) is not yet available in the literature.

The literature on the derivation of macroscopic equations from an underlying microscopic description is limited to models derived according to the structure (7.5.1) in the relatively simpler case of cells which do not have the ability to generate new cells in a population different from the one of the interacting pairs.

A relatively more detailed description can be obtained by assuming that the motion of cells is ruled by forces exchanged by cells described by interaction potentials. The difference with respect to classical mechanics is that these potentials depend on the activity variable. The modeling process shows a remarkable analogy with the modeling of swarms, which will be discussed in the next chapter. The paper by Bellouquid and Delitala (2005) develops a mathematical approach for the modeling of the motion of cells viewed as a swarm.

Development of this class of models can be useful in the challenging perspective of describing the behavior of the overall system by equations derived at different scales, when additional physical events, such as motion under the action of chemoattractants, occur.

8

Modeling Crowds and Swarms: Congested and Panic Flows

8.1 Introduction

The class of models dealt with in this chapter is characterized by complex interactions among active particles that move in two or more space dimensions. The dynamics is remarkably influenced by environmental conditions, as modifications of these conditions can substantially change the rules of interactions. Specifically, the modeling depicts the behavior of crowds and swarms using an approach which includes the ability of individuals to follow specific strategies also generated by their interaction with the outer environment.

The models show a certain analogy with those used for traffic flow modeling in Chapter 6. In that chapter the approach of the kinetic theory for active particles was used to provide a mathematical description of traffic phenomena, characterized by interacting entities, i.e. driver-vehicle systems, with the ability to organize their movement according to a collective, statistically distributed strategy common to all of them.

However, the strategy for crowds and swarms depends on environmental conditions related to situations which, in some cases, change in time; for instance, when panic conditions arise. Moreover, these models have to include the ability of the interacting individuals to organize their dynamics according to a certain ability to *think*, in addition to specific strategies such as collective movement towards a well-defined objective.

Additional complexity aspects of the modeling of crowds and swarms compared to traffic flow modeling include the following.

i) The dynamics takes place in two or three space dimensions, whereas traffic flow is defined in one space dimension and is one directional (although one may also study multilane flow, which requires the modeling of the passage from one lane to the other.

ii) All drivers have approximately the same strategy, which is not consistently modified by outer conditions. Namely, the driver always aims at reaching the end of the trip at the desired velocity within a technical adaptation involving density of vehicles, weather conditions, etc. On the other hand, the strategy of crowds and swarms may differ consistently from individual to individual, as well as with respect to environmental conditions.

iii) The dynamics of the interactions and the overall strategy for crowds and swarm is modified according to specific situations, for instance the presence of panic may change it consistently.

iv) For crowds, the mathematical statement of the problem requires boundary conditions in addition to the inlet and outlet of particles. Moreover, for a swarm, the zone containing it is confined by a boundary which moves in time according to the dynamics of the overall system.

This is a fascinating, however difficult, topic, and it is capturing more and more the attention of applied mathematicians and physicists. However, only some preliminary results are available in the literature. This chapter provides an introduction to the topic, and offers several suggestions and hints to the reader as conceivable research perspectives. The analysis distinguishes between crowds (a system of individuals in an urban environment) and swarms (a system of individuals in a free landscape). The difference is not only technical, but includes several substantial issues to be carefully considered in the modeling process.

Some papers are available in the literature mainly concerning modeling issues at the microscopic and macroscopic scale, among others, Schweitzer (2003), Henderson (1997), Helbing (1992), and Lovas (1994). The recent papers by Hughes (2002), (2003), which deal with macroscopic-type modeling, show clearly that the modeling can be developed only if the *thinking* ability of interacting individuals is carefully taken into account. Indeed, this is the main guiding paradigm of this chapter.

Vicsek (2004) analyzes a strategy to select the scales at which the preceding phenomena can be observed and mathematically described. The motivation in modeling the dynamics of crowds is also related to the control of panic situations or engineering structural safety as documented in the paper by Venuti, Bruno, and Bellomo (2007).

The mathematical literature in the field is not as developed as it is for traffic flow modeling. Therefore, this chapter mainly looks at research perspectives rather than analyzing of well-established mathematical issues. The analysis is mainly referred to modeling crowd dynamics, and the additional technical difficulties in the modeling of swarms are subsequently considered

The chapter is organized into four more sections.

– Section 8.2 provides a mathematical representation of crowds and swarms within the framework of kinetic theory, and also within a continuum mechanics approach at the macroscopic scale. Indeed, the analysis of models at the macroscopic scale contributes to our understanding of the modeling approach using methods of the kinetic theory for active particles.

– Section 8.3 analyzes some modeling aspects at the macroscopic scale showing various technical differences between modeling crowds with respect to swarms. As we shall see, the difference involves not only modeling, but also the statement of the initial and initial-boundary value problems. This section, is preliminary to the following ones devoted to modeling by kinetic theory methods.

– Section 8.4 presents a model of the system under consideration using the mathematical kinetic theory approach for active particles. As we shall see, the modeling must include long distance interactions related to the strategy developed by individuals to organize their dynamics with reference to the presence of other individuals. The interested reader can find guidelines to develop a new modeling approach that are also finalized to the statement of mathematical problems for computational analysis.

– Section 8.5 develops a critical analysis of the preceding sections mainly addressed to indicate research perspectives. The main issue in the modeling is the mathematical description of the strategy developed by the individuals. This strategy technically differs for crowds and swarms and depends both on the collective objective of all particles (individuals) and on their reaction to the surrounding ones. The mathematical description of the interactions among individuals means identifying the strategy they develop, and the analyses at the lower and higher scales may contribute to this objective.

One difference of this chapter with respect to the preceding ones is that the analysis of specific models is not developed here, although various research perspectives are brought to the attention of the reader. This is due to the fact that the literature on the topics under consideration is not well established, although interest in this field is rapidly growing due to its social importance. Therefore, the aim of this chapter is to contribute to research programs concerning crowd and swarm modeling.

8.2 The Representation of Crowds and Swarms

The representation of large systems of interacting individuals organized in crowds and swarms can be obtained by a one-particle distribution function over the microscopic state of the interacting individuals to be regarded as active particles. More than in other cases, the representation must take into account the ***thinking*** ability of the interacting individuals, which plays a relevant role in the interaction and finally in the overall evolution of the system.

This section studies the mathematical representation of the above systems within the frameworks of both kinetic theory and continuum mechanics. As indicated in Section 8.1, the analysis of the macroscopic-type modeling is a useful reference to that of the kinetic theory for active particles, which will be developed in this chapter. As in the preceding chapters, dimensionless variables are used.

Unlike the case of vehicular traffic flow, the literature in the field is quite limited, therefore, an effort is made to unify the notation and methodological approaches. It is hoped that this unification can contribute to improve research activity in the field.

Let us consider a large system of individuals, regarded as active particles, over a two-dimensional domain $\Omega \subseteq \mathbb{R}^2$, which can be either bounded or unbounded. The following quantities can be used for the identification of dimensionless independent and dependent variables.

• ℓ is a characteristic length of the system. If Ω is bounded, then ℓ is the largest dimension; if Ω is unbounded, then ℓ is the largest dimension of the domain containing the initial position of all active particles.

• n_M is the maximum density of the particles corresponding to their maximum admissible packing.

• V_M is the maximum admissible mean velocity which can be reached, on average, by the specific individuals under consideration in free flow conditions, while the maximum admissible velocity for an isolated individual is denoted by $(1 + k)V_M$, where k is a dimensionless parameter.

These quantities allow the assessment of the following independent variables:

• t which is the dimensionless time variable obtained by referring the real time t_r to the critical time $T_c = V_M/\ell$;

• x, y which are the dimensionless space variables obtained by referring the real space variables x_r and y_r to ℓ.

Consider now, using the preceding variables, the representation at the microscopic and macroscopic scales, which means, respectively, identification of the state of each active particle and of the local averages of the microscopic state of the active particles.

Specifically, the microscopic state is defined by the following variables:

- $\mathbf{x} = \{x, y\}$ which is the position in Ω of the individuals;
- $\mathbf{v} = \{v_x, v_y\}$ which is the velocity of the individuals.

The macroscopic representation is defined by the following variables:

$\rho = \rho(t, x, y)$ which is the density referred to the maximum density n_M of individuals;

$\vec{\xi} = \vec{\xi}(t, x, y)$ which is the mean velocity, referred to V_M, written in two space dimensions as

$$\vec{\xi}(t, x, y) = \xi_x(t, x, y)\,\mathbf{i} + \xi_y(t, x, y)\,\mathbf{j}\,, \qquad (8.2.1)$$

where $\mathbf{i}$ and $\mathbf{j}$ denote the unit vectors of the coordinate axes.

Generally, the modeling is limited to conservation and equilibrium for mass and linear momentum, and energy is not taken into account, although some reasoning concerning this specific quantity is developed. The above quantities have to be regarded as the average in space, at a fixed time, of the state of individuals in the elementary area of the space variable. The same quantities can be obtained using moments of a probability distribution function when such a representation is defined according to the methods of generalized kinetic theory.

Let us now consider the problem of representing a system constituted by a large number of interacting individuals regarded as active particles, whose microscopic *activity* variable is identified by a scalar variable $u \in D_u \subseteq \mathbb{R}$, which describes the intensity by which individuals develop their specific strategy.

The state of the whole system is defined by the statistical distribution of position and velocity of the individuals regarded as active particles:

$$f = f(t, \mathbf{x}, \mathbf{v}, u) = f(t, x, y, v_x, v_y, u)\,, \qquad (8.2.2)$$

where $f(t, \mathbf{x}, \mathbf{v}, u)\,d\mathbf{x}\,d\mathbf{v}\,du$ denotes the number of individuals which, at time t, are in the elementary domain of the microscopic states

$$[x, x + dx] \times [y, y + dy] \times [v_x, v_x + dv_x] \times [v_y, v_y + dv_y] \times [u, u + du]\,.$$

The distribution function f can be normalized with respect to n_M so that all derived variables can be given in a dimensionless form. Macroscopic observable quantities can be obtained, under suitable integrability assumptions, by moments of the distribution. In particular, the **dimensionless local density** is given by

$$\rho(t, \mathbf{x}) = \int_{D_\mathbf{v}} \int_{\mathbb{R}_+} f(t, \mathbf{x}, \mathbf{v}, u) \, d\mathbf{v} \, du \,. \qquad (8.2.3)$$

The **total number of individuals** which are in Ω at time t is given by

$$N(t) = \int_{\Omega} \rho(t, \mathbf{x}) \, d\mathbf{x} \,. \qquad (8.2.4)$$

In the same way, the **mean velocity** can be computed as follows:

$$\vec{\xi}(t, \mathbf{x}) = \frac{1}{\rho(t, \mathbf{x})} \int_{D_\mathbf{v}} \int_{\mathbb{R}_+} \mathbf{v} \, f(t, \mathbf{x}, \mathbf{v}, u) \, d\mathbf{v} \, du \,, \qquad (8.2.5)$$

and similarly the **speed variance**,

$$\sigma(t, \mathbf{x}) = \frac{1}{\rho(t, \mathbf{x})} \int_{D_\mathbf{v}} \int_{\mathbb{R}_+} \left(\mathbf{v} - \vec{\xi}(t, \mathbf{x}) \right)^2 f(t, \mathbf{x}, \mathbf{v}, u) \, d\mathbf{v} \, du \,. \qquad (8.2.6)$$

The speed variance provides a measure of the stochastic behavior of the system with respect to the deterministic macroscopic description.

These macroscopic quantities can also be related to the flow $\mathbf{q}$ as follows:

$$\mathbf{q}(t, \mathbf{x}) = \vec{\xi}(t, \mathbf{x}) \, \rho(t, \mathbf{x}) \,, \qquad (8.2.7)$$

where

$$\mathbf{q}(t, \mathbf{x}) = \int_{D_\mathbf{v}} \int_{\mathbb{R}_+} \mathbf{v} \, f(t, \mathbf{x}, \mathbf{v}, u) \, d\mathbf{v} \, du \,, \qquad (8.2.8)$$

which, as in the case of traffic flow, is a quantity which can be measured with a better precision than the mean velocity.

It is plain that the modeling of behavioral differences among individuals is lost in the macroscopic description because the quantities above are obtained by an averaging process, including that over u. The individual dynamics is replaced by the average of that of all individuals.

8.3 Modeling by Macroscopic Equations

This section introduces the derivation of macroscopic equations with the aim of contributing to a deeper understanding of the modeling approach by methods of the generalized kinetic theory for active particles.

The mathematical structures to be used for modeling are formally the same for both crowds and swarms. The approach is based on mass and momentum conservation equations obtained assuming that the flow of particles is continuous with respect to both dependent variables. Generally, suitable smoothness assumptions are needed concerning the behavior of the dependent variable with respect to space. The momentum equation can be replaced by conservation of suitable Riemannian invariants.

The essential differences with respect to traffic flow modeling are that the system has more than one space dimension, and that the acceleration term in the momentum equation must be modeled not only on the basis of the action applied to the elementary mass-volume by the surrounding elements, but also on a general collective strategy of the individuals belonging to the crowd or swarm. Moreover, the acceleration differs substantially in conditions of panic with respect to normal flow conditions.

This aspect also has an influence the kinetic-type modeling considering that the motion of the active particles is also due to external actions. A deeper understanding of the acceleration term in the macroscopic models helps improve the modeling of the microscopic interactions.

The modeling can take advantage of empirical data obtained by experiments. As in traffic flow, measurements are performed only in steady, uniform, flow conditions. But in this case, the data need to be used in unsteady conditions, which produces technical difficulties in the modeling approach.

Moreover, dealing with systems in more than one space dimension not only increases the computational complexity, but because the system may be confined into bounded domains, it also requires a statement of proper boundary conditions that is more complex than those in one space dimension. This issue is discussed later with reference to the modeling approach by kinetic-type equations.

The modeling approach developed here is essentially based on the two cited papers, Hughes (2002) and (2003). To reduce the notation, it is presented in two space dimensions.

Bearing all this in mind, let us first consider the formal statement of the equations for mass conservation and momentum equilibrium. They can be stated, referring to the dimensionless variables defined in Section 8.2, as follows:

$$
\begin{cases}
\dfrac{\partial \rho}{\partial t} + \dfrac{\partial}{\partial x}(\rho \xi_x) + \dfrac{\partial}{\partial y}(\rho \xi_y) = 0, \\[2ex]
\dfrac{\partial \xi_x}{\partial t} + \xi_x \dfrac{\partial \xi_x}{\partial x} = \mathcal{A}_x\big[\rho, \vec{\xi}\,\big], \\[2ex]
\dfrac{\partial \xi_y}{\partial t} + \xi_y \dfrac{\partial \xi_y}{\partial y} = \mathcal{A}_y\big[\rho, \vec{\xi}\,\big],
\end{cases}
\tag{8.3.1}
$$

where $\mathcal{A}_x$ and $\mathcal{A}_y$ model the component of the mean acceleration.

The modeling problem consists in the mathematical description of the accelerations F_x and F_y which depend on the local flow conditions. Consistently with the macroscopic approach only averaged quantities are taken into account. Some examples of acceleration models are given, and their validity will be critically analyzed with specific reference to crowds and swarms. The acceleration refers to the elementary (fictitious) mass of individuals in the area $[\mathbf{x}, \mathbf{x} + d\mathbf{x}]$ around the point $P = \{x, y\}$.

Bearing all this in mind, the acceleration can be viewed as the superposition of two contributions due to adaptation to the mean flow velocity measured under steady uniform flow conditions ξ_e and to local density gradients. Both contributions are assumed to act along the unit vector $\vec{\nu} = \vec{\nu}(x, y)$ directed from $P(x, y)$ to the target $T(x_T, y_T)$ as shown in Figure 8.3.1.

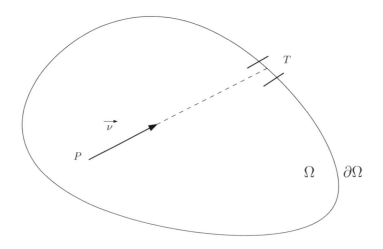

Fig. 8.3.1: Geometry of the space domain.

The modeling can be improved by including a mathematical description of the ability of individuals of the crowd to select paths along lines of minimum density rather than along the straight line identified by $\vec{\nu}$. However, the presentation here is limited to the above-mentioned first level approximation; the analysis of macroscopic-type models is used simply to borrow some ideas useful to the modeling approach of mathematical kinetic theory. The interested reader is addressed to Hughes (2002) and (2003) for a deeper understanding of modeling by macroscopic-type equations.

The calculation of the unit vector $\vec{\nu}$, according to the geometry of the system, is as follows:

$$\vec{\nu}(x,y) = \vec{\nu}_x(x,y) + \vec{\nu}_y(x,y), \qquad (8.3.2)$$

where

$$\vec{\nu}_x(x,y) = \frac{x - x_T}{\sqrt{(x - x_T)^2 + (y - y_T)^2}}\, \mathbf{i}\,,$$

and

$$\vec{\nu}_y(x,y) = \frac{y - y_T}{\sqrt{(x - x_T)^2 + (y - y_T)^2}}\, \mathbf{j}\,,$$

where the direction of the vector is simply identified, with obvious meaning of the notation, by the coordinates of the points $P = P(x,y)$ and $T = T(x_T, y_T)$.

The modeling of the velocity ξ_e can be developed by borrowing some ideas from macroscopic traffic flow modeling. Specifically, with reference to Bonzani and Mussone (2003) as well as to the general overview in the book by Kerner (2004), it is shown that there exists a critical density ρ_c such that ξ_e keeps its maximum value $\xi_e = 1$ for all densities smaller or equal to ρ_c, while ξ_e decays monotonically, with increasing density, to the value $\xi_e = 0$ corresponding to the maximum density $\rho = 1$. The above-cited paper proposes the following model:

$$\begin{cases} \rho \leq \rho_c : & \xi_e = 1, \\[2mm] \rho \geq \rho_c : & \xi_e = \exp\left\{-\alpha \dfrac{\rho - \rho_c}{1 - \rho}\right\}, \end{cases} \qquad (8.3.3)$$

where α is a parameter which takes into account the overall quality of the environment including weather and visibility conditions. Low values of α correspond to good quality, where the decay of velocity with density is not very steep, whereas the slope of the decay increases in bad conditions corresponding to higher values of α. The paper by Venuti and Bruno (2007) reports empirical data specifically related to crowd dynamics and proposes a technical modification of (8.3.3), that appears consistent with experimental data on crowd dynamics.

The modeling of the acceleration can be stated as follows:

$$\vec{A}[\rho; x, y] = \mathcal{A}_T[\rho]\,\vec{v}(x,y) + \mathcal{A}_G[\rho]\,\vec{v}(x,y),\tag{8.3.4}$$

where the first term corresponds to the adaptation to the velocity $\vec{\xi}_e$ and the second one to the influence of local density gradients.

For instance, the following simple models can be used:

$$\mathcal{A}_T = c_T\left(\xi_e(\rho) - \xi\right),\tag{8.3.5}$$

and

$$\mathcal{A}_G = -c_G\frac{\partial\rho}{\partial s},\tag{8.3.6}$$

where c_T and c_G are constants and s is a scalar coordinate along the direction of $\vec{v}$, and where $\xi = |\vec{\xi}|$.

Suppose that the density ρ is a smooth function of the space everywhere. Then the mathematical model is written as follows:

$$\begin{cases}
\dfrac{\partial\rho}{\partial t} + \dfrac{\partial}{\partial x}(\rho\xi_x) + \dfrac{\partial}{\partial y}(\rho\xi_y) = 0, \\[2mm]
\dfrac{\partial\xi_x}{\partial t} + \xi_x\dfrac{\partial\xi_x}{\partial x} = c_T\left(\xi_e(\rho) - \xi\right)\dfrac{x - x_T}{\sqrt{(x-x_T)^2 + (y-y_T)^2}} - c_G\dfrac{\partial\rho}{\partial x}\,v_x, \\[4mm]
\dfrac{\partial\xi_y}{\partial t} + \xi_y\dfrac{\partial\xi_y}{\partial y} = c_T\left(\xi_e(\rho) - \xi\right)\dfrac{y - y_T}{\sqrt{(x-x_T)^2 + (y-y_T)^2}} - c_G\dfrac{\partial\rho}{\partial y}\,v_y.
\end{cases}$$

$$\tag{8.3.7}$$

This modeling can be improved in various ways. For instance, we can replace ρ by a fictitious density (assumed by to be felt by the individuals), where the real density is increased in the case of positive density gradients and decreased in the case of negative gradients. An additional improvement corresponds to the ability of individuals to select trajectories with low density gradients rather than the straight line from the points P to T.

This brief description is sufficient to derive some useful information for modeling by methods of kinetic theory. The preceding analysis refers to modeling crowd dynamics; some technical differences immediately appear in the modeling of swarms. This topic is discussed in the next section referring to models developed by methods of kinetic theory. Some qualitative indications are listed here.

i) The dynamics of swarms is in three space coordinates, while the one of crowds can be approximated in two space coordinates.

ii) Mathematical problems are stated in unbounded domains with initial conditions with compact support. Problem solutions should provide the evolution in time of the domain of the initial conditions.

iii) Accelerations of the fictitious elementary mass depend on its localization and it varies from the border to the center of the swarm.

8.4 Modeling by Kinetic Theory Methods

The analysis developed in the preceding sections has been based on a continuum mechanics approach. Modeling at the macroscopic scale contributes, however, to a deeper understanding of the modeling approach by methods of the mathematical kinetic theory for active particles. Therefore, after the *excursus* of the preceding section, the analysis turns again to the aims of the book.

The first step in modeling using kinetic theory methods is the selection of the appropriate structure to be used for the derivation of specific models. Therefore, a detailed analysis both of the overall common strategy developed by the crowd and of the interactions among individuals, generates specific models.

Similarly to the case of macroscopic equations one has to consider, for the test individuals, both the objective, assumed to be common to all individuals, to reach a zone (or a point) of the space domain, and the interactions with other individuals. The latter modify the dynamics related to the first objective.

The class of systems analyzed in this section is characterized by the ability of particles (individuals) to look around, within a certain visibility zone, and adjust their velocity to the flow conditions around them. Modification of the velocity occurs, at least in the absence of panic, with negligible influence of short range interactions between particles. This suggests that we use the framework corresponding to long range interactions rather than to localized interactions. The reference framework can be found in Section 2.4 of Chapter 2.

Bearing all this in mind, consider the dynamics in two space dimensions in a domain $\Omega : \mathbf{x} \in \Omega \subseteq \mathbb{R}^2$. If Ω is bounded, then its boundary is indicated by $\partial \Lambda$. Moreover, consider first the relatively simple case when the activity variable is the same for all particles. Then the distribution function can be written as follows:

$$f(t, \mathbf{x}, \mathbf{v}) \, \delta(u - u_0), \tag{8.4.1}$$

where the activity variable has been assumed, for simplicity, to be scalar.

Some preliminary assumptions are needed to develop the modeling process.

Assumption 8.4.1. *The test active particle in $\{x, y\}$ is subject to an acceleration $\mathbf{F}$ directed along $\vec{\nu}$ with an intensity F depending on the macroscopic density conditions:*

$$\mathbf{F} = F[\rho]\,\vec{\nu}(x, y)\,, \tag{8.4.2}$$

where $\vec{\nu}$ identifies the direction towards the target objective as in Figure 8.3.1.

Assumption 8.4.2. *The test active particle is subject to long distance interactions, which depend on the position and activity of the interacting pairs, described by the vector*

$$\mathcal{P} = \mathcal{P}(x, x_*, y, y_*)\,. \tag{8.4.3}$$

Therefore, the vector acceleration due to interactions is

$$\mathcal{F}[f](t, \mathbf{x}, \mathbf{v}) = \int_D \mathcal{P}(\mathbf{x}, \mathbf{x}_*, \mathbf{v}, \mathbf{v}_*) f(t, \mathbf{x}_*, \mathbf{v}_*)\, d\mathbf{x}_*\, d\mathbf{v}_*\,, \tag{8.4.4}$$

where $D = \Lambda \times D_{\mathbf{v}}$, and where Λ is the interaction domain of the test active particle:

$$\mathbf{x}_* \notin \Omega \Rightarrow \mathcal{P}_{ij} = 0\,. \tag{8.4.5}$$

The mathematical framework to be used for modeling is given by the following equation:

$$\frac{\partial}{\partial t} f(t, \mathbf{x}, \mathbf{v}) + \mathbf{v} \cdot \nabla_{\mathbf{x}} f(t, \mathbf{x}, \mathbf{v})$$

$$+ \nabla_{\mathbf{v}} \cdot \left(\left(\mathbf{F}[\rho] + \mathcal{F}[f] \right) f(t, \mathbf{x}, \mathbf{v}) \right) = 0\,. \tag{8.4.6}$$

The modeling problem consists then in the mathematical description of the above macroscopic and microscopic accelerations, respectively $\mathbf{F}[\rho]$ and $\mathcal{F}[f]$, depending on various conceivable physical situations.

The first term can be modeled by borrowing some ideas from macroscopic modeling. Consequently, using the same notation:

$$\mathbf{F}[\rho] = \left[\alpha_1 \left(\xi_e(\rho) - \xi \right) - \alpha_2 \frac{\partial \rho}{\partial s} \right] \vec{\nu}(x, y)\,, \tag{8.4.7}$$

where α_1 and α_2 denote suitable parameters.

Let us now consider, referring to the second term, the unit vector $\vec{\mu}(\mathbf{x}, \mathbf{x}_*)$ from $\mathbf{x}$ to $\mathbf{x}_*$, and the distance $d(\mathbf{x}, \mathbf{x}_*)$ between $\mathbf{x}$ and $\mathbf{x}_*$. The action can be assumed to be directed along $\vec{\mu}$ and dependent on the distance: repulsive for $d < d_c$ and weakly attractive for $d \geq d_c$. Formally:

$$\mathcal{P} = \beta_1 \left[\frac{d_c - d(\mathbf{x}, \mathbf{x}_*)}{d(\mathbf{x}, \mathbf{x}_*)} \right] \exp \left\{ - \beta_2 \left(d_c - d(\mathbf{x}, \mathbf{x}_*) \right)^2 \right\} \vec{\mu}, \qquad (8.4.8)$$

where β_1 and β_2 are suitable parameters of the model.

Substituting expressions (8.4.7) and (8.4.8) into (8.4.6) yields a very simple model to be regarded as an indication to develop a specific modeling approach.

Some improvements are immediate, for instance:

i) Assuming that the parameters α_1, α_2, β_1, and β_2 depend on the local density;

ii) Adding to the expression of $\mathbf{F}$ the possibility that individuals will to follow lines of minimum density gradients rather than a straight line.

More generally, both models (8.4.7) and (8.4.8) can be properly improved after a detailed analysis of the interactions among individuals in the crowd. The above modeling approach refers to crowd dynamics; some technical differences considered in the modeling of swarms are analyzed in the next section.

Further improvements can be obtained by supposing that the distribution function and the rules of interaction, depend on the activity variable. The simplest case is when such a distribution is constant in time, as then it is not modified by interactions. In this case, one has

$$f(t, \mathbf{x}, \mathbf{v}) \, P_0(u), \qquad (8.4.9)$$

where

$$\int_0^\infty P(u) \, du = 1. \qquad (8.4.10)$$

In this case the parameters α and β can be modeled as depending on u, in the sense that increasing values of u also increase the response of individuals to outer stimuli.

More generally, the distribution depends on u, say $f = f(t, \mathbf{x}, \mathbf{v}, u)$, and the general modeling framework is now

$$\frac{\partial}{\partial t} f(t, \mathbf{x}, \mathbf{v}, u) + \mathbf{v} \cdot \nabla_{\mathbf{x}} f(t, \mathbf{x}, \mathbf{v}, u)$$

$$+ \nabla_{\mathbf{v}} \cdot \left((\mathbf{F}[\rho] + \mathcal{F}[f]) \, f(t, \mathbf{x}, \mathbf{v}, u) \right)$$

$$+ \nabla_u \left(\mathcal{F}[f] f(t, \mathbf{x}, \mathbf{v}, u) \right) = 0. \qquad (8.4.11)$$

Modeling interactions at the microscopic scale should also include a probability distribution over the activity variable and the interactions that modify it. For instance, it can be assumed that interactions between particles with a state close to each other have a trend to approach the reciprocal state, while the opposite behavior occurs when the distance between the two states is sufficiently large. An assumption of this type corresponds to modeling the behavior of individuals to share the same goal, while individuals keen to move fast oblige other individuals to slow down.

8.5 Looking Forward

Various issues concerning modeling crowd dynamics have been analyzed in this chapter, from the approach at the macroscopic scale using equations of continuum mechanics to the one of the mathematical kinetic theory for active particles according to the frameworks of Chapter 2.

The topics in this chapter have been only recently approached by applied mathematicians, and the literature is not as well developed as it is for traffic flow. However, the topics are attracting a growing interest due also to practical motivations, such as the analysis and control of danger or panic situations.

Therefore, this chapter has been mainly addressed to define guidelines for modeling perspectives. In this section, still referring to some analogies (and differences) with respect to traffic flow modeling, we analyze how the approach developed in the preceding sections for crowd dynamics, can be adapted to modeling the dynamics of swarms in an unbounded domain. An additional issue that we analyze is the modeling of panic situations starting from normal ones.

This last section specifically deals with these topics also in connection with the statement of mathematical problems, typically the initial-boundary value problem. This section contain four subsections, concluding with a critical analysis. The analysis refers to kinetic-type modeling, leaving to the initiative of the interested reader the corresponding analysis for models at the macroscopic scale.

8.5.1 On the Statement of Mathematical Problems

The statement of mathematical problems, typically initial value problems in the whole space, and initial-boundary value problems in bounded domains, also contributes to a deeper understanding of modeling aspects of the dynamics of swarms.

The initial value problem for models of crowd dynamics is stated in unbounded domains for active particles (individuals) who have the objective of reaching a point of the whole space. The statement is obtained by linking the model delivered by Eq. (8.4.6) to initial conditions:

$$f_0(\mathbf{x}, \mathbf{v}, u) = f(t = 0, \mathbf{x}, \mathbf{v}, u), \quad \forall \mathbf{x} \in D_{t_0} \subset \mathbb{R}^2, \quad \forall \mathbf{v} \in D_{\mathbf{v}}, \quad (8.5.1)$$

where D_{t_0} is the domain of the space variable where the initial condition is localized. See Figure 8.5.1.

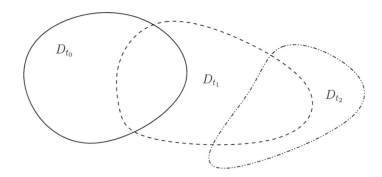

Fig. 8.5.1: Time evolution of the domain occupied by the crowd.

The solution of the problem should provide the evolution D_t of the domain occupied by the particles. The problem is meaningful if the target T does not belong to the domain of the initial conditions and until T is not included in D_{t_0}. Otherwise, the problem needs additional boundary conditions as in the problem which follows.

Let us now consider the initial-boundary value problem for models of crowd dynamics in a fixed domain Ω with boundary $\partial\Omega$. Moreover, let T be a part of the boundary where particles flow out, and $\mathcal{I}$ be the part where

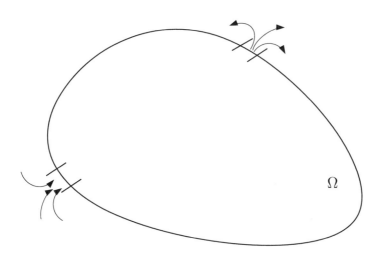

Fig. 8.5.2: Closed domain with inlet and outlet boundaries.

the inlet of individuals occurs. Finally, let us denote by $\partial\Omega_b$ the part of the boundary where no inlet or outlet occurs. See Figure 8.5.2.

The statement of the initial-boundary value problem is then defined by linking Eq. (8.4.6) both to the initial condition (8.5.1) and to boundary conditions for each of the boundary domains $\mathcal{T}$, $\mathcal{I}$, and Ω_b.

Suppose that the boundary $\partial\Omega$ is regular and that the vector $\mathbf{n}$, directed towards the inside of the space domain, can be defined everywhere on $\partial\Omega$. Then the distribution functions of the active particles (individuals) going towards $\partial\Omega$ or leaving $\partial\Omega$ can be denoted, respectively, by

$$f_i = f(t, \mathbf{x} \in \partial\Omega, \mathbf{v}_i, u), \tag{8.5.2}$$

where $\mathbf{v}_i$ is velocity towards the wall defined by the condition $\mathbf{v} \cdot \mathbf{n} < 0$ and

$$f_r = f(t, \mathbf{x} \in \partial\Omega, \mathbf{v}_r, u), \tag{8.5.3}$$

where $\mathbf{v}_r$ denotes the velocity leaving the wall identified by the condition $\mathbf{v} \cdot \mathbf{n} \geq 0$.

Boundary conditions on Ω_b can be stated as follows:

i) For $\mathbf{x} \in \mathcal{I}$ the distribution function of the particles leaving the surface is prescribed for all times:

$$f(t, \mathbf{x} \in \mathcal{I}, \mathbf{v}, u), \quad \forall t \geq 0. \tag{8.5.4}$$

ii) For $\mathbf{x} \in \Omega_b$ the map $f_i \to f_r$ is prescribed for all times as follows:

$$f_r = \mathcal{K} f_i, \quad \forall t \geq 0, \tag{8.5.5}$$

where $\mathcal{K}$ is a linear integral operator which, borrowing some ideas from mathematical kinetic theory, can be detailed as follows:

$$f_r(t, \mathbf{x} \in \Omega_b, \mathbf{v}_r, u) = \int_{D_{\mathbf{v}_i}} \mathcal{K}(\mathbf{v}_i \to \mathbf{v}_r; u) f_i(t, \mathbf{x} \in \partial\Omega_b, \mathbf{v}_i, u) \, d\mathbf{v}_i,$$
$$\tag{8.5.6}$$

where

$$\int_{D_{\mathbf{v}_r}} \mathcal{K}(\mathbf{v}_i \to \mathbf{v}_r; u) \, d\mathbf{v}_r = \int_{D_{\mathbf{v}_i}} \mathcal{K}(\mathbf{v}_i \to \mathbf{v}_r; u) \, d\mathbf{v}_i = 1, \tag{8.5.7}$$

with $\mathcal{K}$ positive defined for all values of its arguments.

iii) For $\mathbf{x} \in \mathcal{T}$ the map $f_i \to f_r$ needs to be properly modeled to describe the flow out of particles:

$$f(t, \mathbf{x} \in \mathcal{T}, \mathbf{v}, u), \quad \forall t \geq 0. \tag{8.5.8}$$

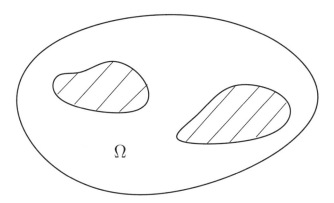

Fig. 8.5.3: Closed domain with internal obstacles.

Figure 8.5.3 shows that a closed domain includes internal obstacles. In this case the statement of the problem should also impose reflection laws on the surfaces of the internal obstacles.

8.5.2 On the Modeling of Swarms

The modeling of swarms is an attractive research perspective, motivated, e.g., by the observation of the beauty of the shapes formed by birds in the sky during spring and autumn periods. Analogous phenomena are, however, observed in other systems such as fishes trying to escape the attack of a predator, or cells which aggregate forming particular shapes.

The statement of the mathematical problem is an initial value problem with data in D_{t_0}, while the domain evolves in time.

Generally, a swarm does not have a precise target to reach unless it has to escape a danger signal appearing from the outer environment. However, one expects that paths of minimum density gradients are followed, while interactions at the microscopic scale follow rules somehow similar to those of crowds. Moreover, the behavior on the border of the swarm follows rules different from those of particles well inside the swarm.

Bearing all this in mind, some guidelines are given for a modeling project also taking into account the indications given in Section 8.3.

 i) Interactions between active particles of a swarm are in three space coordinates, while those of particles of a crowd are defined over two space coordinates.

 ii) Mathematical problems are stated in unbounded domains with initial conditions with compact support. The problem solution should provide the evolution in time of the domain of the initial conditions.

iii) The test particle is not subject to acceleration directed towards a precise objective, and modeling of microscopic interactions should also take into account the localization of the interacting pairs. The output of the interactions is different at the border than in the center of the swarm.

These indications should be regarded as a very preliminary step towards the development of models suitable to describe the complex dynamics of the system under consideration.

8.5.3 Transition from Regular to Panic Conditions

The behavior of crowds and swarms in panic situations appears to be substantially different from that observed under normal flow conditions. In some cases, behaviors under panic conditions generate extremely dangerous situations which may possibly be avoided by control of individual behaviors. This subsection is devoted to understanding the various technical issues involved in the modeling of panic situations.

The analysis refers to models derived in the framework of the mathematical kinetic theory for both classes of models: crowds and swarms. Both the collective strategy and the individual interactions have to be considered. Again, some qualitative indications for research perspectives can be given to model panic conditions:

i) Test particles are not subject to an acceleration directed towards a precise objective, because they follow, generally without a practical motivation, the other particles.

ii) Pair interactions in panic conditions substantially differ from those in normal flow conditions as attracting accelerations become dominant with respect to repulsive ones. Moreover, their intensity becomes relatively greater.

iii) The individual behavior in panic conditions tends to be the same for all individuals as all of them become aware of the environmental situation and become active. Technically this means that the distribution over the activity variable tends to a Dirac delta function over high values of the activity variable.

Again, these indications offer some very preliminary suggestions towards research programs in this attractive field.

8.5.4 Critical Analysis

Various modeling aspects of crowds and swarms have been analyzed in this chapter, including the description of panic situations, which may possibly modify the rules by which active particles interact among themselves.

The literature in the field is not yet settled because the strategy for the modeling problems is not yet well defined. However, applied mathematicians are strongly attracted by these topics, so we expected that in the very near future interesting contributions will be added and a proper modeling strategy defined.

This chapter has also provided a brief introduction to modeling at the macroscopic scale, with the idea that modeling by kinetic equations can benefit from a knowledge of modeling by hydrodynamic-type equations. Indeed, such a modeling approach, although questionable considering that the assumption of continuity of matter does not appear to be fully consistent, contributes to a deeper understanding of the collective strategy of the interacting particles.

A few more ideas are proposed in this subsection in an attempt to define a modeling project using methods of mathematical kinetic theory.

• The mathematical structure to be used in the modeling is the one corresponding to long range interactions. The role of the activity also needs to be included in modeling interactions because it plays an important role in describing transitions from normal to panic situations.

• The mathematical structure proposed in this chapter should be regarded as a general paradigm to derive new, more refined models. The improvement should refer to a detailed modeling of microscopic interactions.

• The analysis of crowd dynamics may include the attempt of individuals (active particles) to escape danger signals generated within the fixed domain where they move. Therefore, the analysis can be addressed to understanding the substantial modifications in the outlet flow to escape the danger. Possibly, the evolution from normal to panic situations, which is fast, but not instantaneous, should be described by the model.

• The analysis of the swarm dynamics may include the attempt of the swarm to escape danger signals generated outside the evolving domain of the localization of the swarm. Therefore, the analysis can be addressed to understanding the substantial modifications in the evolution of the domain. Again, the fast evolution from normal to panic situations should be described by the model.

• Discretization of the velocity variable or of the activity, using methods somehow analogous to those we have seen for traffic flow models, is also meaningful in the modeling of crowds and swarms to tackle the problem of the assumption of continuity of the probability distribution over the microscopic state which is not consistent when the number of interacting particles is small (in some sense still to be properly defined).

The above suggestions should be regarded as a (limited) panorama. The interested reader can discover new perspectives. In all cases, the analysis of models needs to be related to the qualitative and computational analysis of mathematical problems, an additional challenging research perspective.

9

Additional Concepts on the Modeling of Living Systems

9.1 Introduction

The preceding chapters were devoted to the development and application of the methods of the mathematical kinetic theory for active particles to the modeling of large living systems of interacting entities.

The term **active particles** has been used to denote entities whose microscopic state includes, in addition to geometrical and mechanical variables (classically position and velocity), a further variable, called **activity**, which represents the ability of the interacting particles to organize their behavior according to individual strategies typical of the system under consideration. Interactions modify the whole microscopic state by a dynamics, where laws of classical mechanics can be modified by the presence of the activity variable.

The overall state of the system is described by the probability distribution function over the microscopic state, while the macroscopic description is obtained by weighted moments of the distribution function. A detailed analysis of microscopic interactions and of the strategies developed by the active particles leads, using technical calculations, to an evolution equation (the model) for the distribution function.

The contents of the preceding chapters can be divided into two parts. The first one was devoted to the methodological aspects of determining various classes of equations to be used as paradigms for the derivation of specific models; the second part to the analysis of specific classes of models concerning social dynamics, vehicular traffic flow, immune competition, and the dynamics of crowds and swarms.

189

The selection of these physical systems is related to the author's personal experience. However, the aim was to select a limited number of examples to show the practical application of the method to a variety of cases and provide a background for future developments.

Indeed, the applications refer to well-defined different structural properties of the models. Specifically:

• Models of social dynamics refer to systems without space structure and with a constant number of active particles. The mathematical structures refer to the evolution of the distribution function over the activity variable only, while space and velocity are not considered. Moments of order zero, corresponding to the number of individuals, and of order one, to the total wealth, are preserved during the evolution.

• Traffic flow models are characterized by space heterogeneity, while the number of active particles (driver-vehicles) is constant in time in the absence of inlet and outlet flows. Therefore, the moment of order zero, corresponding to the number of vehicles, is constant, while the moment of order one, referred to linear momentum, changes in time and space due to different flow conditions. The inlet and outlet of vehicles leads to a variable number of active particles, however their number is known in time.

• Complex biological multicellular systems are characterized by proliferative and/or destructive events which modify the number of interacting individuals. Moreover, interactions may produce active particles in a new population different from that of the interacting pairs, so that the number both of interacting active particles and populations may grow in time.

• Models of crowds and swarms differ from those of the preceding chapters by the presence of a collective strategy in addition to the individual one. Moreover, such a strategy can be modified by environmental situations, for instance, by the presence of collective panic situations. The dynamics is in two or more space dimensions, while the number of particles is variable in time due to inlet and outlet of individuals.

The different structural properties of the mathematical frameworks used to derive these models correspond to their well-defined physical differences. A proper selection of the background mathematical framework is strategic towards the whole modeling process.

This final chapter proposes some additional concepts on the modeling of living systems with regard to the approach of the kinetic theory for active particles. We provide a critical overview of the difficulty in representing living systems by mathematical equations and look at the perspective idea of developing a proper mathematical theory for living systems analogous to some of the celebrated theories in mathematical physics.

It is a fascinating perspective, however difficult, and may require years to be achieved even for a few specific cases. Nevertheless, it is worth trying, and the difficulty of the may be problem an incentive for mathematicians.

The content of this chapter is as follows.

– Section 9.2 provides an overview of mathematical problems generated by the application of models to the study of real world phenomena. Specific topics are the statement and analysis of initial and initial-boundary value problems for isolated systems and for interconnected networks of systems.

– Section 9.3 analyzes some conceivable developments of the mathematical structures focused on multiple interactions, interconnected systems of systems, and open systems. This section also introduces hybrid systems, where models corresponding to different scales may be present in the overall model.

– Section 9.4 briefly looks at modeling perspectives related to other types of systems, among others epidemiology, pair formation, opinion formation, and communication networks. The aim is to show how the methodological approach can be properly generalized for a variety of modeling objectives.

– Section 9.5 finally looks at the perspective idea that a mathematical theory can be developed for specific classes of living systems with some analogy to the classical approach of mathematical physics. The natural starting point for this challenging, and fascinating, perspective, is a critical overview of the complexity of the system to be analyzed.

The contents of this chapter are focused mainly on research perspectives. Unlike the preceding chapters, here an effort is made to reduce the formalization as far as possible, leaving appropriate developments to future research activity.

9.2 Mathematical Problems

The application of models to the qualitative and computational analysis of real world systems generates mathematical problems which are of great interest, due partly to their technical difficulty, to applied mathematicians. Typically, initial value and initial-boundary value problems are considered.

The *initial value problem* is stated by implementing the evolution equations, which may be formally written in the continuous case as follows:

$$\mathcal{L}f = \mathcal{N}f, \tag{9.2.1}$$

where $f = f(t, \mathbf{x}, \mathbf{v}, \mathbf{u})$. With suitable initial conditions,

$$f_0(\mathbf{x}, \mathbf{v}, \mathbf{u}) = f(t = 0, \mathbf{x}, \mathbf{v}, \mathbf{u}), \quad \forall \, \mathbf{x}, \mathbf{v} \in D_{\mathbf{x}} \times D_{\mathbf{v}}. \tag{9.2.2}$$

The generalization of this statement, valid for one population only, to systems of equations and to models with discrete states is immediate.

The statement can be used in the spatially homogeneous case, for instance referred to models of Chapter 5, or to vehicular traffic flow models (Chapter 6) depending on space, but with periodic boundary conditions, e.g., when the road is a ring.

An interesting case is the qualitative analysis of problems with initial conditions given on a compact closed domain and a vacuum outside. In this case, the solution of the problem should also provide the time evolution of the boundary, considering that the velocity of active particles is always bounded.

When active particles are contained in a bounded domain, say for $\mathbf{x} \in \Gamma \subset \mathbb{R}^3$ (or in the case of external flows), the statement of initial-boundary value problems requires that we link Eq. (9.2.1) to suitable boundary conditions for $\mathbf{x} \in \partial\Gamma$, where $\partial\Gamma$ is the boundary of Γ and $\vec{\nu}$ is the unit vector orthogonal to $\partial\Gamma$ and directed from the wall. Active particles may modify their microscopic state, due to interaction with the walls, while the surface can allow inlet or outlet flows.

The statement of boundary conditions is obtained by splitting the distribution function f into a part f^- corresponding to particles which approach the wall $\mathbf{v} \cdot \vec{\nu} < 0$, and a part f^+ corresponding to particles which leave the wall $\mathbf{v} \cdot \vec{\nu} > 0$, respectively:

$$f^- = f(t, \mathbf{x} \in \partial\Gamma, \mathbf{v} \cdot \vec{\nu} \leq 0, \mathbf{u}), \tag{9.2.3a}$$

and

$$f^+ = f(t, \mathbf{x} \in \partial\Gamma, \mathbf{v} \cdot \vec{\nu} > 0, \mathbf{u}). \tag{9.2.3b}$$

The statement of the problem is then obtained by linking Eq. (9.2.1) to the initial conditions (9.2.2) and to boundary conditions defined by the functional relation

$$f^+ = \mathcal{R} f^-, \tag{9.2.4}$$

where $\mathcal{R}$ is a suitable operator mapping f_- into f^+. The difference with respect to the statement of boundary conditions for classical particles, summarized in Chapter 1, is that interactions of active particles with the wall also depend on the activity variable.

The above statement can be made relatively less general by assuming that interactions are localized in space and are number conservative. In this case, one can assume that the map has the same structure defined in Chapters 1 and 8, whereas when walls allow a flux of particles, either positive or negative, this source term has to be added.

A simplified description can be obtained by assuming that interactions with the walls are localized, and the map (9.2.4) can be specialized as follows:

$$f(t, \mathbf{x} \in \partial\Gamma, \mathbf{v}^+ \cdot \vec{\nu} \leq 0, \mathbf{u}^+) = \int_{(D_\mathbf{v} \times D_\mathbf{u})} c|\mathbf{v}^- \cdot \vec{\nu}|\mathcal{M}^w(\mathbf{v}^- \rightarrow \mathbf{v}^+|\mathbf{u}^-)$$

$$\times \mathcal{B}^w(\mathbf{u}^- \rightarrow \mathbf{u}^+) f(t, \mathbf{x} \in \partial\Gamma, \mathbf{v}_- \cdot \vec{\nu} > 0, \mathbf{u}^-) \, d\mathbf{v}^- \, d\mathbf{u}^-, \qquad (9.2.5)$$

where the plus and minus signs correspond, respectively, to $\mathbf{v} \cdot \vec{\nu} \geq 0$ and $\mathbf{v} \cdot \vec{\nu} \leq 0$. Moreover:

• $\mathcal{M}^w(\mathbf{v}^- \rightarrow \mathbf{v}^+|\mathbf{u}^-)$ denotes the probability density, conditioned by $\mathbf{u}^-$, that a particle with velocity $\mathbf{v}^-$ will be reflected with velocity $\mathbf{v}^+$;

and

• $\mathcal{B}^w(\mathbf{u}^- \rightarrow \mathbf{u}^+)$ denotes the probability density that a particle reaching the wall with activity $\mathbf{u}^-$ will be reflected with activity $\mathbf{u}^+$.

Similarly to models of particle interactions, the terms $\mathcal{M}$ and $\mathcal{B}$, corresponding to the interactions with the walls, should be modeled according to a phenomenological interpretation of the physical behavior of the real system.

Applied mathematicians are attracted by this class of problems due to their complexity related to the lack of conservation equations; they are used to derive *a priori* estimates necessary to analyze long time behavior. Referring again to Chapters 5–7: models of social dynamics can take advantage of conservation of mass and linear momentum, models of traffic flow can exploit only mass conservation, and the presence of proliferative and destructive events does not allow us to use mass conservation for biological systems.

The literature for classical models of kinetic theory proposes a variety of interesting mathematical results as documented in the books by Cercignani, Illner, and Pulvirenti (1994) and Glassey (1995), and more recently in the review papers by Perthame (2004) and Villani (2004). The analysis of the links between models for classical and active particles is developed in the paper by Arlotti, Bellomo, and De Angelis (2002).

The analysis of the initial value problem in the kinetic theory for active particles is mainly limited to the spatially homogeneous case, as documented in the papers by Bertotti and Delitala (2004) and (2007) for models of social dynamics, by De Angelis and Jabin (2003) and (2005), and by Bellouquid and Delitala (2004) and (2006) for different models of immune competition.

Some of these results can be critically analyzed with regard to their contribution to a deeper understanding of the predictive ability of the model.

Indeed, the validation of models is generally based on their ability to predict interesting phenomena at a qualitative level, rather than organizing a comparison with empirical data, which are not always available and reliable. For instance, this is the case for models of social dynamics, where the analysis can be addressed to identify how a certain, politically induced, social dynamics may lead (or not) to a desired distribution of social wealth. Similarly, the analysis of immune competition can be focused to identify the role of biological parameters on the output of the competition. Some details are given in the following paragraphs.

• The qualitative and computational analysis by Bertotti and Delitala (2007) focuses on the above issue and has shown that the solution to the initial value problem tends, asymptotically in time, to a stable equilibrium configuration which depends on the wealth distributed among the social classes and on the social dynamics. As mentioned in Chapter 5, the authors achieve the above results, working on suitable Lyapunov functionals, which operate only for discrete state models with a low number of social states.

Computational experiments show that the above qualitative behavior is the same when we increase the number of social states. Therefore, an open problem is to prove asymptotic stability for an arbitrarily large number of social states, possibly also in the case of continuous approximation. It may be that computational empirical evidence is sufficient for the applications, however this mathematical problem is sufficiently tricky to attract the attention of mathematicians.

An additional problem is the analysis of several interconnected systems, perhaps nations or regions of a nation. The whole may be regarded as a system of systems. A qualitative analysis may show (which is still to be proven) how the asymptotic trend differs in each system due to the reciprocal action of the interconnected ones.

• The methodological approach also applies to the analysis of equilibrium configurations of traffic flow models in the spatially homogeneous case. This type of analysis has been developed, for two technically different discrete velocity models, by Coscia, Delitala, and Frasca (2007) and by Delitala and Tosin (2007).

The mathematical stability result is obtained only for a low number of velocities; it is still to be proven for an arbitrarily large number of velocities. Similarly to the case of social dynamics, an open problem is the analysis of several interconnected roads regarded as systems so that the whole is a system of systems. The qualitative analysis may show (still to be proven) how the asymptotic trend differs in each road due to the reciprocal action of the interconnected ones.

• The qualitative and computational analysis of models in the presence of proliferative and/or destructive events has been developed for models such as those in Chapter 7 for multicellular systems and immune competition. A variety of results is reported in the book by Bellouquid and Delitala (2006) for models which do not describe shifting from one population to the other. The remarkable difficulty encountered in the qualitative analysis is the lack of conservation equations.

The objective of the qualitative analysis is the study of the trend of solutions and, in some cases, of the asymptotic behavior in time. Considering that models describe the competition between progressing and immune cells, the analysis is required to investigate whether or not immune cells are able to suppress the presence of progressing cells. Further, it should focus on the role of the parameters of the model in the competition.

The interested reader is addressed to the above-cited book where several interesting results are reported corresponding to different types of competition. Various theorems concerning the trend of solutions are obtained by linking the analysis of the evolution equations for the distributions to those for the number densities obtained as first-order moments. A detailed use of sharp inequalities give the result for the number densities, and a computational analysis completes the description by showing the corresponding evolution for the whole distribution function.

In principle, a similar analysis can be developed for models which include shifts from one population to the other by daughter cells generated into a population different from those of the interacting pair. It is an interesting and still open problem. A deeper understanding of this complex system may possibly enlighten the phenomena of the evolution of progression and heterogeneity of malignant cells, see Komarova (2006).

Analogous indications can be given for models depending on space. Various computational experiments are available for different models, however a detailed qualitative analysis of the initial value problem is not yet available.

• The statement of initial-boundary value problems has been given above. The contents of the preceding chapters has indicated various problems where a qualitative and computational analysis is required to provide a description of interesting physical phenomena. A typical example is the one of a crowd confined in a bounded domain with individuals obliged to interact with walls undergoing a dynamics conditioned by the presence of obstacles. A different problem is posed by the analysis of swarms where initial conditions are given in a closed domain with a vacuum outside. The analysis should provide the evolution of the domain with free boundaries. It is a rather delicate problem considering that the dynamics depends also on the shape of the boundary.

This overview refers only to a few examples out of a variety of interesting problems posed by the application of models to real world phenomena. All of them are regarded as conceivable research perspectives because the existing literature is limited to some computational experiments which still lack a detailed qualitative analysis.

9.3 Looking for New Mathematical Structures

The common feature of the examples of mathematical models in the preceding chapters is that the real system is constituted by a large number of interacting entities, called active particles, which have the ability to express functions addressed to organize their dynamics according to individual and collective strategies.

The main objective of the modeling is to describe the collective behavior starting from models of individual behavior and pair interactions. The mathematical structures offered by Chapters 2–4 have been used to derive different classes of models. The last section of the various modeling chapters has critically analyzed the strategy to select a certain structure over another: for instance, a discrete state model rather than a continuous one; or short range rather than long range interactions, depending on the inner structure of the real system selected for the modeling.

It has been shown that the strategy to select one framework rather than another depends on which kind of *prediction* or *exploration* is provided by the model. Therefore, it is useful to speculate on the probability of finding improvements or even alternative structures to those we have seen in the preceding chapters. This critical analysis is developed in this section by attempting to give at least a partial, response to the questions stated below.

 i) Do alternatives to the mathematical approach developed in Chapters 2–4 exist?

 ii) Can the mathematical structures proposed in this book be simplified to achieve a relatively more immediate description of physical reality?

iii) Can the structures be improved to achieve a deeper mathematical description of living systems, at the cost of a higher complexity of the model?

We will keep the analysis induced by the above questions at the level of qualitative indications, implemented by bibliographic indications, with the aim of focusing future research activity.

• **Modeling alternatives:** The answer to the first question is definitely positive. It would be too naive to state the opposite. Looking for a mathematical approach (and tools) to model living systems is a challenge which will engage the intellectual energies of applied mathematicians in the coming years.

Although some mathematical methods, including those developed here, appear to be promising (and in some cases successful) in modeling certain specific systems, an exhaustive result has not been achieved. The mathematical approach should be able to deal with a broad variety of systems.

The methods of this book claim to be an alternative to the traditional approach where the specific peculiarities of living systems are not captured. That is due to the fact that the main ingredients of modeling consist of deterministic causality principles. On the other hand, our alternative may be only the beginning of the story.

In particular, the interested reader can find in the book by Schweitzer (2003) several interesting issues (and a bibliography) on the mathematical approach by master equation methods, while transport equation methods with special attention to biological systems are discussed in the book by Perthame (2007).

• **Simplification of the mathematical structures:** Let us now consider the second question. In the first chapter, it was stated that hoping to handle the complexity of living systems by mathematical structures simpler than those needed to deal with inert matter is a contradiction. On the other hand, reducing the complexity of models is always a useful objective as long as the predictive or explorative ability of the model is not lost.

It may be argued that looking for a structural simplification is an interesting objective when the main functions expressed by the living system are not suppressed. Further, the functions are not the same for all individuals. In other words, the expressions of the functions are generally distributed over the individuals and evolve in time.

Applying this reasoning to the search for simple structures implies that population-type models where the state of the system is identified by the number of individuals belonging to a certain population do not respond to the modeling requirement that the activity functions are not the same for all individuals. These models are stated in terms of ordinary differential equations and provide a description at a super-macroscopic scale where individuals of a certain population are regarded as a whole, all of them with the same state, and are simply identified by their number.

We can improve the descriptive ability of population dynamics models by introducing internal variables, such as delay, maturation, and biological functions, which are the same for all entities homogeneously distributed in space. In this case models are stated in terms of partial differential equations taking the partial derivative with respect to the internal structure. See, among several works, Rotenberg (1983), Webb (1986), Perthame and Ryzhik (2005).

An alternative that deserves attention is the mathematical approach proposed by Cattani and Ciancio (2007). They suggest considering population dynamics models with random coefficients linked to a probability density which evolves in time due to interactions at a microscopic level.

The evolution is described by mathematical structures similar to those of Chapter 2 for systems that are spatially homogeneous or with a vanishing role of the space and velocity variables. This approach may be a candidate for the mathematical description of some specific living systems. This idea can be properly exploited to derive models at the macroscopic scale with parameters delivered by the dynamics at the lower scale.

• **Further generalization of the mathematical structures:** Finally, let us consider the third question concerning further enrichment of the mathematical structures. As usual the *dilemma* is that any enrichment involves an increasing complexity both in the assessment of the model and in the development of computational schemes for its analysis.

Considering that the above dilemma cannot be analyzed in full generality, but only for specific cases, some developments can be considered leaving the strategy to use them or not only when related to well-defined applications. Of course, one can consider improvements as only substantial modifications of the mathematical structures rather than technical generalizations such as those we have seen in the preceding chapters.

The following issues are brought to the attention of the reader: modeling multiple interactions, and modeling interconnected systems. Referring to the first issue, we observe that all structures and models in this book are based on the assumption that only binary interactions play a relevant role in the modeling. However, triple interactions must play a role in the analysis of living systems, and one has to analyze how much these interactions influence the evolution of the system.

In some cases, multiple interactions can play a remarkable role, which is a hint to deal with the mathematics of multiple interactions. This issue has already generated interesting mathematical problems concerning game theory with more than two players, see Platkowski (2004), and the qualitative analysis of the solutions to the Cauchy problem, see Arlotti and Bellomo (1996), is developed for systems in which the microscopic state does not include space and velocity variables.

A recent paper by Bellomo and Carbonaro (2006) has developed methods of the mathematical kinetic theory for active particles to model the evolution of complex psychological games and reciprocal feelings in the case of multiple interactions. The analysis is limited to binary and triple interactions and is referred to the movie *Jules and Jim*, directed by Francois Truffaut. The story involves three persons: Jeanne Moreau (Cathérine), Oskar Werner (Jules), and Henri Serre (Jim) in a complex interaction over a long period which includes the first world war.

The analysis of these papers has shown how the methods of the mathematical kinetic theory for active particles can also be used to model the formation and evolution of personal feelings involving three persons. Moreover, the analysis has been addressed to understand how involving triple collisions greatly increases the complexity of the mathematical approach.

It has been shown that enlarging the number of interacting populations (or individuals regarded as a population) does not substantially increase the complexity with the only exception that one equation is needed for each population. Moreover, the modeling can be technically developed using a scalar variable to identify the microscopic state within each population. This strategy consistently reduces the complexity of the modeling at the microscopic scale, which is essential for deriving specific models referring to the mathematical structures of the kinetic theory for active particles.

However, when multiple interactions are involved, the assessment of the microscopic state, called activity, can no longer be identified by a scalar variable. This means that the number of populations is smaller than the dimension of the variable needed to identify the whole set of microscopic states. That is a remarkable source of complexity.

Technically, the activity variables should be referred to all specific interchanges of feelings. In the case of the above-cited paper, each person, identified as a population, is the carrier of three feelings. This entrains that the whole microscopic state must be represented by a matrix with issues related to all encounters including self-feelings derived from personal meditations.

Modeling interconnected systems means dealing with different interacting models, whose difference may be simply in the parameters or even in their mathematical structure. An example of the first case has been analyzed in Chapter 5 (the model of social dynamics of several interacting nations), while interactions of models with different structures generally occur when a model derived at a certain scale interacts with a model derived at a different scale. This occurs for biological systems when the overall model attempts to describe interactions of phenomena at the cellular scale with what happens at the molecular (subcellular scale).

A different way to approach the modeling process is to look at the overall system as an assembly of several subsystems, each one expressing a specific activity function. The situation and the mathematical difficulty are analogous to those outlined above. In addition, suitable compatibility conditions have to be worked out in the connections linking the parts of the whole system.

9.4 Additional Issues on Modeling

As mentioned in Section 9.1, the mathematical models in Chapters 4–8 have been selected according to the author's specific experience. On the other hand, the reader interested in modeling developments should have received sufficient information to develop a modeling approach to complex systems different from those analyzed in the preceding chapters.

The selection of mathematical structures plays an essential role in the modeling process, followed by the proper statement of mathematical problems. In general, models should refer to the structures developed in Chapters 2–4. The selection of the applications has also been strategically referred to the idea that different structures correspond to different classes of models.

The use of models with discrete or continuous activity variables mainly corresponds to identification problems, but not to mathematical structures. In general the same model may refer either to a continuous structure of Chapters 2–3, or to the corresponding discrete structure of Chapter 4.

This section briefly outlines some conceivable additional applications referred to those we have seen in the preceding chapters. These indications are given without mathematical formalizations, leaving their developments to well-defined research programs. The bibliography reported in the following paragraphs mostly refers to the master equation approach for Brownian agents, see Schweitzer (2003).

• *Models of social dynamics* (Chapter 5) can be properly developed to deal with economical and social sciences viewed as complex evolutionary systems, see Arthur, Durlauf, and Lane (1997), and Axelrod (1997). Generalizations are directly related to the specific interpretation of the meaning of active particles and activity which may be quite different from those considered in the examples of Chapter 5.

In particular, and with some analogy with complex biological systems, the active particles may not be individuals or social classes, but modules with the ability to collectively express certain functions, say enterprises, state institutions, or governments, which are regarded as interacting populations. Consequently, the activity variable has a different meaning for each module. The literature in the field reports several interesting contributions concerning general issues and the master equation or Brownian agents approach, among others: Helbing (1995), Kirman and Zimmermann (2001), Schweitzer (2003).

Various fields of social sciences can be analyzed according to this approach, e.g., opinion formation, see among others Schweitzer, Zimmermann, and Mühlenbein (2002), or organization of decision, see, e.g., Schweitzer and Holyst (2002). Opinion formation may develop under the action of media, where the first population is constituted by individuals who are receptors of the *opinion* coming from other populations, for instance, political parties, who attempt to transmit, by different means, a certain opinion.

Some physical systems consider social behaviors in connection with aggregation phenomena with pattern formation, see (among others) Koch and Meinhardt (1994), and Mogilner, Edelstein-Keshet, Bent, and Spiros (2003). Modeling is often developed without considering the heterogeneous distribution of the activity variable, while such a role may be relevant as documented in Chapters 3 and 5 of the book by Schweitzer (2003).

Some mathematical approaches, technically referred to developments of the methods of kinetic theory, make use of the model introduced by Smoluchowski to describe coagulation and fragmentation phenomena, see the review by Wattis (2006). This class of equations, which has been developed by various authors, e.g., Fasano, Rosso, and Mancini (2006), either with continuous or discrete states, shows a structure analogous to the one proposed in this book. Again, this structure should be generalized including the role of heterogeneous distribution of the activity variable. A conceivable application refers to urban aggregation.

A systematic approach using methods of the kinetic theory for active particles has been recently initiated, Ajmone Marsan and Bellomo (2008), to consider most of the above social phenomena.

• *Models of vehicular traffic, and of the dynamics of crowds and swarms* (Chapters 6 and 8) must be properly developed by including the role of the activity variable to model the different random behavior of each driver-vehicle system. Some indications have been given in the chapters. This development appears to be a necessary research perspective for enabling models to describe the behavioral differences among individuals.

Bearing all this in mind, additional developments can be indicated. For instance, the modeling of networks is an interesting perspective which has already captured the attention of applied mathematicians, e.g., Schweitzer, Ebeling, Rosè, and Weiss (1998), Schweitzer and Tilch (2002). Mathematical theory and optimization problems have been dealt with by Coclite, Garavello, and Piccoli (2005), and by Garavello and Piccoli (2006). The unsolved problem still appears to be implementing the traffic flow in the interconnected lanes by realistic models, while at present the computational complexity has suggested that we deal with very simple models derived at the macroscopic level.

A quite natural development appears to be the analysis of pedestrian flows in urban roads; see, among others, Ebeling and Schweitzer (2002), Helbing (1992), and Helbing, Schweitzer, Keltsch, and Molnár (1997). The analysis should take into account the modeling both of crowds and traffic in networks. The main problem is again linking the interconnected paths with an accurate modeling of the pedestrian flows within them.

Particularly interesting is the analysis of panic situations, which modify substantially the behavior of pedestrians and, in particular, the rules by which they interact. Some indications have been given in Chapter 8 towards a research perspective which is definitely worth developing. The above reasoning can be extended to the modeling of swarms, where the main difficulty still appears to be modeling interactions at the microscopic level, Czirok and Vicsek (2000).

Traffic on networks does not simply refer to vehicles and pedestrians, but also to a variety of flows such as the internet: Garetto, Lo Cigno, Meo, and Ajmone Marsan (2004), Ajmone Marsan, Garetto, Giaccone, Leonardi, Schiattarella, and Tarello (2005), Ajmone Marsan, Leonardi, and Neri (2005), or vascular blood flow in angiogenesis phenomena: Folkman (2002), Drasdo (2005). Modeling may take advantage of dedicated computational devices such as cellular automata: Deutsch and Dormann (2005).

• *Models of complex systems in biology* deserve special attention because the scientific community appears convinced that a great revolution of this century is going to be the mathematical formalization of phenomena pertaining to biological sciences. Thus the heuristic experimental approach, which is the traditional investigative method in biological sciences, should be gradually linked to new methods and paradigms generated by a deep interaction with mathematical sciences, Woese (2004).

A modeling approach analogous to the one of Chapter 7 can be developed in other fields of the life sciences; e.g., to the modeling of epidemics, where the evolution of the pathology may be related to genetic mutations of the virus. The definition of microscopic scale technically differs for each of the above systems. Specifically, the microscopic scale for the first system corresponds to the cellular scale, and it is the scale of individuals (persons, animals, plants, etc.). The corresponding submicroscopic scale differs again: genes in the first case, cells and virus particles in the second case.

An essential role is delivered by recent studies in genetics and genomics documented in the paper by Vogelstein and Kinzler (2004), which provides a deep qualitative interpretation of phenomena at the molecular scale. The analysis of this paper ought to be properly put into mathematical equations, possibly using evolutionary game theory, Nowak and Sigmund (2004). A further interesting approach, still based on evolutionary game theory, has been proposed by Gatenby, Vincent, and Gillies (2005), focused on the immune competition.

Referring to multicellular systems and introducing some concepts of the next section, we stress that if robust biological theory can provide a mathematical description of cellular interactions, ruled by biological events at the subcellular scale, then this information can possibly be exploited to develop a proper mathematical theory for biological sciences as analyzed by Bellomo and Forni (2006). This may be the most challenging and attractive research perspective.

9.5 Speculations on a Mathematical Theory for Living Systems

The models reported until now including the additional examples in Section 9.4 can be derived by referring to the mathematical structures developed in Chapters 2–4. But these models cannot yet be classified at the level of a proper mathematical theory. This is due to the fact that their derivation has been obtained by implementing the mathematical structures by a description of microscopic interactions obtained by a phenomenological interpretation of physical reality rather than by a proper theory valid in a broad variety of physical situations.

The search for a mathematical theory for living systems definitely deserves some additional reasoning because in the future such a theory will possibly be developed, at least for some specific systems. The above analysis is developed through the following three sequential steps, each of which is discussed in the subsections which follow.

 i) Analysis of the conceptual difficulties in dealing with living systems from the viewpoint of mathematical sciences;

 ii) Methodological approach to develop a mathematical theory of living systems;

iii) Multiscale aspects of a mathematical theory for living systems.

The analysis of the third issue precisely refers to understanding what is still missing in a mathematical theory for living systems. Indeed, this topic appears to be the natural conclusion of this book.

9.5.1 Why is Mathematics for Living Systems so Difficult?

The title of this subsection is very similar to the one of an interesting paper by Reed (2004), who identifies a variety of difficulties faced in the attempt to describe living systems by mathematical equations. The main difficulty appears to be the absence of the invariance principles which exist in the case of inert matter. This aspect was analyzed in Chapter 1 referring not only to the above-cited paper, but also to the one by Hartwell, Hopfield, Leibner, and Murray (1999). We generalize to living systems the reasoning focused on biological systems.

• Although living systems obey the laws of physics and chemistry, the notion of function or purpose differentiates them from systems of other natural sciences. Indeed, what really distinguishes living systems from physics are survival and reproduction, but mainly the concomitant notion of function, which has the ability to modify the equilibrium and conservation laws of classical mechanics.

• Large living systems are different from the physical or chemical systems analyzed by statistical mechanics or hydrodynamics. Statistical mechanics typically deals with systems containing many copies of a few interacting components, whereas some living systems, e.g., those of biology, may contain from millions to a few copies of each of thousands of different components (populations), each (active particle) with specific functions and an ability to interact with the entities of different components.

• Living systems cannot always be simply observed and interpreted at a macroscopic level. A system constituted by millions of interacting entities (active particles) shows at the macroscopic level only the output of the cooperative and organized behaviors which cannot be, or are not, singularly observed.

It follows that a mathematical theory not only must be developed at a well-defined observation and representation scale, but also must show consistency with the whole set of scales which represent the system. In other words, models developed at the microscopic scale (in our case by methods of the mathematical kinetic theory for active particles) are generated by a detailed analysis of models at the microscopic scale, and they should provide, by means of suitable asymptotic theories, a mathematical description at the macroscopic scale.

9.5.2 Methodological Approach

Developing a research project to tackle the above-mentioned conceptual difficulties means that the paradigms of the traditional approach, applied to inert matter, should be replaced by new ones, while applied mathematicians should not attempt to describe complex living systems by simple paradigms and equations.

Indeed, applied mathematicians traditionally have attempted to describe living matter by equations relatively simpler than those used to describe inert matter. Simple equations have generated various unsuccessful attempts and ultimately they constitute an obstacle to the development of a biological-mathematical theory. Applied mathematicians understand well the pessimistic attitude by Wigner (1960), although this is contrasted by the positive contributions of various papers which have appeared in this century.

The traditional approach of mathematical sciences generates ***mathematical models*** which we distinguish from a mathematical theory. Some of them may occasionally reproduce specific aspects of biological phenomena, but rarely are they able to capture the essential features of the biological phenomena.

Although the idea of developing a proper mathematical theory for living systems still appears to be a distant prospect, it is possible. We outline a few guidelines.

• A *preliminary step* in the development of a theory is the selection of the specific classes of systems which must be described by mathematical equations. The situation shows some analogy with the one of mathematical physics, where a theory, although based on almost universal principles, does not apply to all physical systems.

– Considering that this book refers to large systems of several interacting entities in different fields of the life sciences, the development of a theory must, at least, be referred to each specific field.

• The *subsequent step* is the selection of the representation scale identified for the observation and mathematical description of the system. The approach developed in this book primarily refers to the *microscopic scale*, which looks at each component of the system, whose functions are determined by the components at an even lower scale. The microscopic state of each entity (active particle) is a peculiarity of the specific system under consideration. Subsequently, the representation of the overall state is assumed to be delivered by the *statistical distribution over the state at the microscopic level*.

Again a reference to the physical sciences provides some useful background.

– Classical kinetic theory refers to the analysis of the evolution, far from equilibrium, of multiparticle systems. The state of each microscopic entity (the particle) is identified by geometrical and mechanical variables: position and orientation, as well as velocity and rotation. Position and velocity are sufficient to identify the state of particles which may be approximated by point masses.

– On the other hand, the main difficulty in dealing with living systems is that the state of active particles includes activity functions, which differ from population to population and which have the ability to organize their dynamics. Moreover, destructive and reproductive events should be taken into account.

• The *third step* is the derivation of mathematical equations to describe the evolution of the statistical distribution over the microscopic state of the interacting active particles. This objective is mathematically pursued by conservation equations in the elementary volume of the state space: the increase of particles in this volume, i.e., of the number of particles with a certain microscopic state, is the output of interactions which modify both the mechanical and activity states, and include reproduction and destruction of particles.

– The mathematical kinetics of classical particles operates in the same way. The space of the microscopic states of the particles, geometrical and mechanical variables, is called *phase space*, and interactions obey laws of classical mechanics without reproduction and destruction events.

– A remarkable difficulty, which is the relevant one, is that interactions of classical particles obey robust theories of physics, specifically Newtonian mechanics, while in the case of living systems a theoretical approach is not yet available for all types of interactions.

9.5.3 Multiscale Problems Towards a Mathematical Theory

The system of equations mentioned in the third step provides ***mathematical structures*** consistent with the equilibrium and conservation equations used to derive them. There is a rigorous framework for the derivation of models, when a mathematical description of interactions at the microscopic level can be derived by the interpretation of empirical data. However, when the above interactions are delivered by a theoretical interpretation delivered within the framework of living sciences, then we may talk about a ***mathematical theory for living systems***.

In other words, the mathematical framework (including future developments) can be a useful reference for the design of specific models, considering that:

– mathematical models cannot be designed on the basis of a purely heuristic approach. They should be referred to well-defined mathematical structures, which may act as a mathematical theory.

Although it can be claimed that mathematical background structures have been developed, we want to analyze which type of additional work is needed to obtain a proper mathematical theory for living systems. The crucial passage is the modeling of the terms describing microscopic interactions as functions of the microscopic states. Chapters 5–8 have shown how these terms can be modeled according to a phenomenological interpretation of the behavior of the active particles of each specific system. However, this type of information should be obtained by a proper theory consistent with living sciences. When, and only if, this target is effectively reached, then the link between mathematics and living sciences will be complete.

The above reasoning can be referred to the multiscale structure of the systems under consideration. Analogously to the classical kinetic theory of particles, the interaction functions should be obtained by a theoretical analysis of the structure of the particles at the lower scale.

For instance, referring to biological systems, an interesting perspective consists in using models at the subcellular scale to derive biological functions defined at the cellular scale. Possibly, suitable developments of methods of the kinetic theory of active particles at the lower scale could be used to determine the distribution of the evolution of entities at the subcellular level: expression and transfer of genes. Therefore, these expressions may identify biological functions. Analogous ideas can be developed for other classes of living systems. However, if the target is generated by the interaction of two different scientific environments: biology and mathematics, then we are talking about a bio-mathematical theory.

Developing a mathematical theory appears especially difficult, perhaps even impossible, in cases such as traffic flow modeling or evolution of personal feelings, where mathematics interacts with very specific aspects of human behaviors. In this case, mathematical equations can only provide a very rough approximation of systems characterized by too great a complexity to be cast into a mathematical framework.

In addition, a mathematical theory must be consistent with the macroscopic scale. Namely, models at the microscopic scale should provide the macroscopic description that is effectively observed. Consistency with the higher scale also means deriving macroscopic equations, generally described by partial differential systems. However, this derivation requires a relatively more sophisticated mathematical analysis, while most macroscopic models, such as those we have seen in the models of crowds and swarms, are derived by a purely heuristic approach.

Macroscopic equations have to be obtained, from the underlying microscopic description, by letting distances between particles go to zero so that the condensed phase can be mathematically described. The derivation also must consider the case of active particles which may reproduce and grow. Then, macroscopic equations refer to systems with variable mass.

The analysis by Bellomo and Bellouquid (2006) has shown that a relevant role is played by the time scaling, i.e., by the relative rates characterizing the three different phenomena described by the model: space dynamics, biological transitions, and growth phenomena. Indeed, living systems may show different types of interactions which may simultaneously occur with different rates so that even the scaling may change of type time. Different mathematical approaches, which put in evidence the same behavior, are due to various authors. Among others Othmer, Dunbar, and Alt (1988), Othmer and Hillen (2002), Stevens (2004), Perthame (2004), Lachowicz (2005), Filbet, Laurencot, and Perthame (2005), Chalub, Dolak-Struss, Markowich, Oeltz, Schmeiser, and Soref (2006),

The above issue is an interesting research perspective that is approached in the literature only at a very preliminary level. It is one of the many challenging objectives posed by the mathematical attempt to describe living systems.

Collective Bibliography

Ajmone Marsan G. and ***Bellomo N.*** (2008), to appear. Towards a mathematical theory of complex socio-economical systems—Multiscale representation and functional subsystems, *Mathematical Models and Methods in Applied Sciences*, **18**.

Ajmone Marsan M., Garetto M., Giaccone P., Leonardi E., Schiattarella E., and ***Tarello A.*** (2005), Using partial differential equations to model TCP mice and elephants in large IP networks, *IEEE Transactions on Networking*, **13**, 1289–1294.

Ajmone Marsan M., Leonardi E., and ***Neri F.*** (2005), On the stability of isolated and interconnected input-queueing switches under multiclass systems, *IEEE Transactions on Information Theory*, **51**, 1167–1174.

Arlotti L. and ***Bellomo N.*** (1996), Solution of a new class of nonlinear kinetic models of population dynamics, *Applied Mathematical Letters*, **9**, 65–70.

Arlotti L., Bellomo N., and ***De Angelis E.*** (2002), Generalized kinetic models in applied sciences, *Mathematical Models and Methods in Applied Sciences*, **12**, 403–433.

Arthur W.B. (1993), On designing economic agents that behave like human agents, *J. Evolutionary Economics*, **3**, 1–22.

Arthur W.B., Durlauf S.N., and ***Lane D.***, Eds. (1997), **Economy as an Evolving Complex System II**, Addison-Wesley, New York.

Asselmeyer T., Ebeling W., and ***Rosé H.*** (1997), Evolutionary strategies of optimization, *Physical Review E.*, **56**, 1171–1180.

Aw A. and ***Rascle M.*** (2000), Resurrection of "second order" models of traffic flow, *SIAM J. Applied Mathematics*, **60**, 916–938.

Axelrod R. (1997), **The Complexity of Cooperation: Agent Based Models of Competition and Cooperation**, Princeton University Press, Princeton, NJ.

Batty M. and *Xie Y.* (1994), From cells to cities, *Environmental Planning*, **21**, 31–48.

Bellomo N., Ed. (1995), **Lecture Notes on the Mathematical Theory of the Boltzmann Equation**, World Scientific, London, Singapore.

Bellomo N. and *Bellouquid A.* (2006), On the onset of nonlinearity for diffusion models of binary mixtures of biological materials by asymptotic analysis, *International J. Nonlinear Mechanics*, **41**, 281–293.

Bellomo N. and *Carbonaro B.* (2006), On the modeling of complex sociopsychological systems with some reasoning about Kate, Jules and Jim, *Differential Equations and Nonlinear Mechanics*, Article ID 86816, 1–26.

Bellomo N. and *Forni G.* (1994), Dynamics of tumor interaction with the host immune system, *Mathematical and Computer Modeling*, **20**, 107–122.

Bellomo N. and *Forni G.* (2006), Looking for new paradigms towards a biological-mathematical theory of complex multicellular systems, *Mathematical Models and Methods in Applied Sciences*, **16**, 1001–1029.

Bellomo N. and *Gatignol R.*, Eds. (2003), **Lecture Notes on the Discretization of the Boltzmann Equation**, World Scientific, London, Singapore.

Bellomo N. and *Gustafsson T.* (1991), The discrete Boltzmann equation: A review of the mathematical aspects of the initial and initial boundary value problems, *Review Mathematical Physics*, **3**, 137–162.

Bellomo N., Lachowicz M., Polewczak J., and *Toscani G.* (1991), **Mathematical Topics in Nonlinear Kinetic Theory II—The Enskog Equation**, World Scientific, London, Singapore.

Bellomo N. and *Lo Schiavo M.* (2000), **Lectures Notes on the Generalized Boltzmann Equation**, World Scientific, London, Singapore.

Bellomo N. and *Preziosi L.* (1995), **Modeling Mathematical Methods and Scientific Computation**, CRC Press, Boca Raton, FL.

Bellomo N. and *Pulvirenti M.*, Eds. (2000), **Modeling in Applied Sciences: A Kinetic Theory Approach**, Birkhäuser, Boston.

Bellouquid A. and *Delitala M.* (2004), Kinetic (cellular) approach to models of cell progression and competition with the immune system, *Z. Angew. Mathematical Physics*, **55**, 295–317.

Bellouquid A. and *Delitala M.* (2005), Mathematical methods and tools of kinetic theory towards modeling complex biological systems, *Mathematical Models and Methods in Applied Sciences*, **15**, 1639–1666.

Bellouquid A. and *Delitala M.* (2006), **Modeling Complex Biological Systems—A Kinetic Theory Approach**, Birkhäuser, Boston.

Bertotti M.G. and *Delitala M.* (2004), From discrete kinetic theory and stochastic game theory to modeling complex systems in applied sciences, *Mathematical Models and Methods in Applied Sciences*, **14**, 1061–1084.

Bertotti M.G. and *Delitala M.* (2007), Conservation laws and asymptotic behavior of a model of social dynamics, *Nonlinear Analysis RWA*, doi: 10.1016/j.nonrwa.2006.09.012.

Blankenstein T. (2005), The role of tumor stroma in the interaction between tumor and immune system, *Current Opinion Immunology*, **17**, 180–186.

Bonzani I. and *Mussone L.* (2003), From experiments to hydrodynamic traffic flow models. Modeling and parameter identification, *Mathematical and Computer Modeling*, **37**, 1435–1442.

Brazzoli I. (2007), On the discrete kinetic theory for active particles—Mathematical tools for open systems, *Applied Mathematical Letters*, doi: 10.1016/j.aml.2007.02.018.

Brazzoli I. and *Chauviere A.* (2006), On the discrete kinetic theory for active particles modeling the immune competition, *Computational and Mathematical Methods in Medicine*, **7**, 143–158.

Blume-Jensen P. and *Hunter T.* (2001), Oncogenic kinase signaling, *Nature*, **411**, 355–365.

Cabannes H. (1980), **The Discrete Boltzmann Equation (Theory and Applications)**, Lecture Notes University of California, Berkeley.

Capasso V. and *Bakstein V.* (2005), **An Introduction to Continuous TimeStochastic Processes**, Birkhäuser, Boston.

Cattani C. and *Ciancio A.* (2007), A hybrid two scale mathematical tool for active particles modeling complex systems with learning hiding dynamics, *Mathematical Models and Methods in Applied Sciences*, **17**, 171–188.

Cercignani C. and *Gabetta E.* (2007), **Transport Phenomena and Kinetic Theory**, Birkhäuser, Boston.

Cercignani C., *Illner R.*, and *Pulvirenti M.* (1994), the Mathematical theory of Dilated Gases, Springer, Heidelberg.

Chalub F., Dolak-Struss Y., Markowich P., Oeltz D., Schmeiser C., and *Soref A.* (2006), Model hierarchies for cell aggregation by chemotaxis, *Mathematical Models and Methods in Applied Sciences*, **16**, 1173–1198.

Chauviere A. and ***Brazzoli I.*** (2006), On the discrete kinetic theory for active particles—Mathematical tools, *Mathematical Computer Modeling*, **43**, 933–944.

Coclite G., Garavello M., and ***Piccoli B.*** (2005), Traffic flow on a road networks, *SIAM J. Mathematical Analysis*, **36**, 1862–1886.

Coscia V., Delitala M., and ***Frasca P.*** (2007), On the mathematical theory of vehicular traffic flow models II. Discrete velocity kinetic models, *International J. Nonlinear Mechanics*, **42**, 411–421.

Czirok A. and ***Vicsek T.*** (2000), Collective behavior of interacting self-propelled particles, *Physica A*, **281**, 17–29.

Daganzo C. (1995), Requiem for second order fluid approximations of traffic flow, *Transportation Research B*, **29B**, 277–286.

Darbha S. and ***Rajagopal K.R.*** (2002), A limit collection of dynamical systems. An application to model the flow of traffic, *Mathematical Models and Methods in Applied Sciences*, **12**, 1381–1399.

De Angelis E. and ***Jabin P.E.*** (2003), Qualitative analysis of a mean field model of tumor-immune system competition, *Mathematical Models and Methods in Applied Sciences*, **13**, 187–206.

De Angelis E. and ***Jabin P.E.*** (2005), Qualitative analysis of a mean field model of tumor-immune system competition, *Mathematical Methods in Applied Sciences*, **28**, 2061–2083.

Degond P., Pareschi L., and ***Russo G.*** (2003), **Modeling and Computational Methods for Kinetic Equations**, Birkhäuser, Boston.

De Lillo S., Salvatori C., and ***Bellomo N.*** (2007), Mathematical tools of the kinetic theory of active particles with some reasoning on the modeling progression and heterogeneity, *Mathematical and Computer Modeling*, **45**, 564–578.

Delitala M. (2003), Nonlinear models of vehicular traffic flow—new frameworks of the mathematical kinetic theory, *C.R. Mecanique*, **331**, 817–822.

Delitala M. and ***Forni G.*** (2007), From the mathematical kinetic theory of active particles to modeling genetic mutations and immune competition, Internal Report, Politecnico Torino.

Delitala M. and ***Tosin A.*** (2007), Mathematical modeling of vehicular traffic: A discrete kinetic theory approach, *Mathematical Models and Methods in Applied Sciences*, **17**, 901–932.

Derbel L. (2004), Analysis of a new model for tumor-immune system competition including long time scale effects, *Mathematical Models and Methods in Applied Sciences*, **14**, 1657–1682.

Deutsch A. and *Dormann S.* (2005), **Cellular Automaton Modeling of Biological Systems**, Birkhäuser, Boston.

Drasdo D. (2005), Emergence in regulatory networks in simulated evolutionary processes, *Advances in Complex Systems*, **8**, 285–318.

Dunn G.P., Bruce A.T., Ikeda H., Old L.J., and *Schreiber R.D.* (2002), Cancer immunoediting: From immunosurveillance to tumor escape, *Nature Immunology*, **3**, 991–998.

Ebeling W. and *Schweitzer F.* (2002), Swarms of particle agents with harmonic interactions, *Theory Biosciences*, **120**, 207–224.

Fasano A., Rosso F., and *Mancini A.* (2006), Implementation of a fragmentation coagulation scattering model for the dynamics of stirred liquid-liquid dispersion, *Physica D*, **222**, 141–158.

Ferziger J.H. and *Kaper H.G.* (1972), **Mathematical Theory of Transport Processes in Gases**, North-Holland, Amsterdam.

Filbet P., Laurencot P., and *Perthame B.* (2005), Derivation of hyperbolic models for chemosensitive movement, *J. Mathematical Biology*, **50**, 189–207.

Folkman J. (2002), Role of angiogenesis in tumor growth and metastasis, *Seminar Oncology*, **29**, 15–18.

Fowler A.C. (1997), **Mathematical Models in the Applied Sciences**, Cambridge University Press, Cambridge.

Frankhauser P. (1994), **La Fractalité des Structures Urbaines**, Anthropos, Paris.

Galam S. (2003), Modeling rumors: The no plane Pentagon French hoax case, *Physica A*, **320**, 571–580.

Galam S. (2004), Sociophysics: A personal testimony, *Physica A*, **336**, 49–55.

Galam S. (2004), Contrarian deterministic effects on opinion dynamics: "The hung elections scenario," *Physica A*, **333**, 453–460.

Garavello G. and *Piccoli B.* (2006), Traffic flow on networks, *Appl. Math. Series*, **1**, American Institute of Mathematical Sciences.

Garetto M., Lo Cigno R., Meo M., and *Ajmone Marsan M.* (2004), Modeling short lived TCP connections with open multiclass queuing networks, *Computer Networks Journal*, **44**, 153–176.

Gatenby R.A., Vincent T.L., and *Gillies R.J.* (2005), Evolutionary dynamics in carcinogenesis, *Mathematical Models and Methods in Applied Sciences*, **15**, 1619–1638.

Gatignol R. (1975), **Théorie Cinétique d'un Gaz à Répartition Discréte des Vitèsses**, Springer Lecture Notes in Physics n. 36.

Gillet É. (2005), À chaque cancer son scénario aléatoire, *La Récherche*, **390**, 73.

Glassey R. (1995), **The Cauchy Problem in Kinetic Theory**, SIAM Publications, Philadelphia.

Greller L., Tobin F., and *Poste G.* (1996), Tumor heterogeneity and progression: Conceptual foundation for modeling, *Invasion and Metastasis*, **16**, 177–208.

Hanahan D. and *Weinberg R.A.* (2000), The hallmarks of cancer, *Cell*, **100**, 57–70.

Hartwell H.L., Hopfield J.J., Leibner S., and *Murray A.W.* (1999), From molecular to modular cell biology, *Nature*, **402**, c47–c52.

Helbing D. (1992), A fluid-dynamic model for the movement of pedestrians, *Complex Systems*, **6**, 391–415.

Helbing D. (1995), **Quantitative Sociodynamics. Stochastic Methods and Models of Social Interaction Processes**, Kluwer, Dordrecht.

Helbing D. (2001), Traffic and related self-driven systems, *Review Modern Physics*, **73**, 1067–1141.

Helbing D., Schweitzer F., Keltsch J., and *Molnár P.* (1997), Active walker model for the formation of human and animal trail systems, *Physical Review E*, **56/3**, 2527–2539.

Henderson L.F. (1997), On the fluid mechanics of human crowd motion, *Transportation Research B*, **8**, 509–515.

Hogeweg P. and *Hesper P.* (1983), The ontogeny of the interaction structure in bumble bee colonies, *Behavioral Ecological Sociobiology*, **12**, 271–283.

Hughes R.L. (2002), A continuum theory of pedestrians, *Transportation Research B*, **36**, 507–535.

Hughes R.L. (2003), The flow of human crowds, *Annual Review of Fluid Mechanics*, **35**, 169–182.

Jager E. and *Segel L.* (1992), On the distribution of dominance in a population of interacting anonymous organisms, *SIAM J. Applied Mathematics*, **52**, 1442–1468.

Kerner B. (2004), **The Physics of Traffic**, Springer, New York, Berlin.

Kirman A. and *Zimmermann J.*, Eds. (2001), **Economics with Heterogeneous Interacting Agents**, Lecture Notes in Economics and Mathematical Systems, n. 503, Springer, Berlin.

Klar A., Kühne R.D., and *Wegener R.* (1996), Mathematical models for vehicular traffic, *Surveys Mathematical Industry*, **6**, 215–239.

Klar A. and *Wegener R.* (1997), Enskog-like kinetic models for vehicular traffic, *J. Statistical Physics*, **87**(1/2), 91–114.

Klar A. and *Wegener R.* (2000), Kinetic traffic flow models, in **Modeling in Applied Sciences: A Kinetic Theory Approach**, Bellomo N. and Pulvirenti M., Eds., Birkhäuser, Boston.

Koch A.J. and *Meinhardt H.* (1994), Biological pattern formation: From basic mechanisms to complex structures, *J. Theoretical Biology*, 1481–1507.

Kolev M. (2005), A mathematical model of cellular immune response to leukemia, *Mathematical and Computer Modeling*, **41**, 1071–1082.

Kolev M., Kozlowska E., and *Lachowicz M.* (2005), Mathematical model of tumor invasion along linear or tubular structures, *Mathematical and Computer Modeling*, **41** (10), 1083–1096.

Komarova N. (2006), Spatial stochastic models for cancer initiation and progression, *Bulletin Mathematical Biology*, **68**, 1573–1599.

Krugman P. (1996), **Self Organizing Economy**, Blackwell, Oxford.

Lachowicz M. (2005), Micro and meso scales of description corresponding to a model of tissue invasion by solid tumors, *Mathematical Models and Methods in Applied Sciences*, **15**, 1667–1684.

Leutzbach W. (1988), **Introduction to the Theory of Traffic Flow**, Springer, New York.

Lin C.C. and *Segel L.A.* (1988), **Mathematics Applied to Deterministic Problems in the Natural Sciences**, SIAM Publications, Philadelphia.

Lollini P.L., Motta S., and *Pappalardo F.* (2006), Modeling tumor immunology, *Mathematical Models and Methods in Applied Sciences*, **16**, 1091–1124.

Lo Schiavo M. (2003), The modeling of political dynamics by generalized kinetic (Boltzmann) models, *Mathematical and Computer Modeling*, **37**, 261–281.

Lo Schiavo M. (2006), A dynamical model of electoral competition, *Mathematical and Computer Modeling*, **43**, 1288–1309.

Lovas G.G. (1994), Modeling and simulation of pedestrian traffic flow, *Transportation Research B*, **28**, 429–443.

May R.M. (2004), Uses and abuses of mathematics in biology, *Science*, **303**, 790–793.

Mogilner A., Edelstein-Keshet L., Bent L., and ***Spiros A.*** (2003), Mutual interactions, potentials, and individual distance in a social aggregation, *J. Mathematical Biology*, **47**, 353–389.

Nelson P. (1995), A kinetic model of vehicular traffic and its associated bimodal equilibrium solution, *Transport Theory Statistical Physics*, **24**, 383–409.

Nowak M.A. and ***Sigmund K.*** (2004), Evolutionary dynamics of biological games, *Science*, **303**, 793–799.

Nowell P.C. (2002), Tumor progression: A brief historical perspective, *Seminars in Cancer Biology* **12**, 261–266.

Othmer H.G., Dunbar S.R., and ***Alt W.*** (1988), Models of dispersed biological systems, *J. Mathematical Biology*, **26**, 263–298.

Othmer H.G. and ***Hillen T.*** (2002), The diffusion limit of transport equations II: Chemotaxis equations, *SIAM J. Applied Mathematics*, **62**, 1222–1250.

Paveri Fontana S.L. (1975), On Boltzmann like treatments for traffic flow, *Transportation Research*, **9**, 225–235.

Perthame B. (2004), Mathematical tools for kinetic equations, *Bulletin American Mathematical Society*, **41**, 205–244.

Perthame B. (2004), PDE models for chemotactic movements: Parabolic, hyperbolic and kinetic, *Applied Mathematics*, **49**, 539–564.

Perthame B. (2007), **Transport Equations in Biology**, Birkhäuser, Basel.

Perthame B. and ***Ryzhik L.*** (2005), Exponential decay for the fragmentation or cell-division equation, *J. Differential Equations*, 155–177.

Platkowski T. (2004), Evolution of populations playing mixed multiplayer games, *Mathematical and Computer Modeling*, **39**, 981–989.

Platkowski T. and ***Illner R.*** (1985), Discrete velocity models of the Boltzmann equation: A survey on the mathematical aspects of the theory, *SIAM Review*, **30**, 213–255.

Portugali J. (2000), **Self-Organization and the City**, Springer, Berlin.

Prigogine I. and ***Herman R.*** (1971), **Kinetic Theory of Vehicular Traffic**, Elsevier, New York.

Reed R. (2004), Why is mathematical biology so hard? *Notices of the American Mathematical Society*, **51**, 338–342.

Romano A., Marasco A., and *Lancellotta R.* (2005), **Continuum Mechanics Using Mathematica, Fundamentals, Applications, and Scientific Computing**, Birkhäuser, Boston.

Rotenberg M. (1983), Transport theory for growing cell dynamics, *J. Theoretical Biology*, **103**, 181–199.

Schweitzer F. (1998), Modeling migration and economic agglomeration with active Brownian particles, *Advances in Complex Systems*, **1**, 11–37.

Schweitzer F. (2003), **Brownian Agents and Active Particles**, Springer, Berlin.

Schweitzer F., Ebeling F., Rosè H., and *Weiss O.* (1998), Optimization of road and networks using evolutionary strategies, *Evolutionary Computation*, **5**, 419–438.

Schweitzer F. and *Holyst J.* (2002), Modeling collective formation by means of active Brownian particles, *European Physics J.*, **15**, 723–732.

Schweitzer F. and *Tilch B.* (2002), Self-assembling of networks in an agent based model, *Physical Review E*, **66**, 026113.

Schweitzer F., Zimmermann J., and *Mühlenbein H.* (2002), Coordination of decisions in a spatial agent model, *Physica A*, **303**, 189–216.

Shang P., Wan M., and *Kama S.* (2007), Fractal nature of traffic data, *Computers and Mathematics with Applications*, **54**, 107–116.

Stevens A. (2004), The derivation of chemotaxis equations as limit dynamics of moderately interacting many-particles systems, *SIAM J. Applied Mathematics*, **61**, 183–212.

Truesdell C. and *Rajagopal K.R.* (2002), **An Introduction to The Mechanics of Fluids**, Birkhäuser, Boston.

Tyagi V., Darbha S., and *Rajagopal K.R.* (2007), A dynamical system approach based on averaging to model the macroscopic flow of freeway traffic, *Nonlinear Analysis: Hybrid Systems*, doi:10.1016/6j.nahs.2006.12.007.

Venuti F., Bruno L., and *Bellomo N.* (2007), Crowd dynamics on a moving platform: Mathematical modeling and application to lively footbridges, *Mathematical and Computer Modeling*, **45**, 252–269.

Venuti F. and *Bruno L.* (2007), An interpretative model of pedestrian fundamental relations, *C.R. Mecanique*, **335**, 194–200.

Vicsek T. (2004), A question of scale, *Nature*, **418**, 131.

Villani C. (2004), Recent advances in the theory and applications of mass transport, *Contemporary Mathematics*, **353**, 95–109.

Vogelstein B. and *Kinzler K.W.* (2004), Cancer genes and the pathways they control, *Nature Medicine*, **10**, 789–799.

Wattis J.A.D. (2006), An introduction to mathematical models of co-agulation fragmentation processes: A discrete deterministic mean field approach, *Physica D*, **222**, 1–20.

Webb G.F. (1986), A model with proliferating cell population with inherited cycle length, *J. Mathematical Biology*, **23**, 169–282.

Weidlich W., (2000), A Systematic Approach to Mathematical Modeling in the Social Sciences, London Harword Academic Publishers.

Weinberg R.A. (2006), **The Biology of Cancer**, Garland Sciences, New York.

Wigner E. (1960), The unreasonable effectiveness of mathematics in the natural sciences, *Communications Pure and Applied Mathematics*, **13**, 1–14.

Woese C.R. (2004), A new biology for a new century, *Microbiology and Molecular Biology Reviews*, **68** (2), 173–186.

Zhang W.-B. (2005), **Differential Equations, Bifurcations, and Chaos in Economics**, World Scientific, London, Singapore.

Index